Site engineering practice

Philip A. Rougier

MCIOB, ACIArb

Construction Press
London and New York

05892445

Construction Press
an imprint of:
Longman Group Limited
Longman House, Burnt Mill, Harlow
Essex CM20 2JE, England
Associated companies throughout the world

Published in the United States of America
by Longman Inc., New York

© Longman Group Limited 1984

First published 1984

British Library Cataloguing in Publication Data

Rougier, Philip A.
   Site engineering practice.
   1. Building sites
   I. Title
   624   TH375

   ISBN 0-582-41236-6

Library of Congress Cataloging in Publication Data

Rougier, Philip A., 1949–
   Site engineering practice.

   Includes index.
   1. Building sites.   2. Building–Superintendence.
I. Title.
TH375.R68 1983   690   83-5305
ISBN 0-582-41236-6

Set in 10/12 pt Linotron 202 Plantin
Printed in Singapore by Selector Printing Co Pte Ltd

D
690
Rou

General Editor: Colin Bassett, BSc, FCIOB, FFB

Formwork, *M.P. Hurst*
Drainage and Sanitation, *R. Payne*

Other related titles

Dictionary of Soil Mechanics and Foundation Engineering, *J. Barker*
Slipform Concrete, *R.G. Batterham*
Building Production and Project Management, *R.A. Burgess and G. White*
Water Installations and Drainage Systems, *F. Hall*
Site Safety, *J. Laney*
Illustrated Dictionary of Building, *P. Marsh*
Site Engineering, *R.W. Murphy*
Project Management and Construction Control, *G. Peters*

Contents

Preface

Most consultants and many senior managers in the construction field fondly remember their early sun-filled days on site – peak physical fitness, fresh qualifications, bright new minds. But they probably prefer to forget the niggling worries, the insecurity, the pressure and uncertainty of having real and direct technical responsibility for construction works on a large scale.

It has been shown many times that the sort of information contained in this book, in the manner presented, is desperately needed particularly by relatively inexperienced site engineers when faced with a situation not dealt with in college lectures. More broadly, the book provides important reference information rarely presented in such a format, over a wide range of situations, of essential interest to the busy practitioner at all levels.

Site Engineering Practice began as an accumulation of tables, notes and sketches kept in a folder and used constantly. It has been developed into a broad compendium of the many subjects and disciplines to which site engineering, as a recognised occupation, owes its roots. Whilst it does not attempt the academic 'treatment' found in so many excellent books, it does give a fully technical and highly practical slant on many topics essential to site practice. Some items are thought to be previously unpublished material, others presented in a greatly simplified form for rapid assimilation and immediate usefulness.

A number of people have been kind enough to assist in the preparation of this book, and thanks are due to Jim Byrne, Julian Collins, Alan Davies and John Threadgold for their technical advice, to Kevin Crook for help with illustrations and to Janet Wheeler who typed the manuscript. The publishers are to be thanked for their patience and consideration during the long period from inception. My wife Elizabeth and daughter Gemma have been a continuous source of encouragement, and without such support the project may well have floundered.

Philip A Rougier

Acknowledgements

Various organisations have supplied valuable data and, where appropriate, have kindly given their permission for use of copyright material. Thanks are due to:

Blandford Plant Hire & Contracting Co Ltd
British Standards Institution
Her Majesty's Stationery Office
Hymac Ltd
James Bros (Hamworthy) Ltd
JCB Sales Ltd
National House-Building Council

Chapter 1

Equipment

Equipment of the correct type and in good working order is a prerequisite for effective performance, so much so that without the right equipment a site engineer is often unable to function at all. It is the equipment that a site engineer uses, and its state of readiness, that distinguishes him from his office-based counterparts. Hours, days or even weeks of precious contract time can be lost if the site engineer is incorrectly equipped.

Most site engineering tasks require the use of an instrument or tool, and in most cases several operations are required, each employing different pieces of equipment. Fortunately, most of the items have a variety of functions, covering a wide range of techniques, but the list is still quite a long one:

Typical equipment list

1 Torpedo level
2 30 m tape/50 m band
3 Claw-hammer
4 Junior hacksaw
5 Nails
6 Level-book
7 200 m line
8 Arrows
9 Retracting (Stanley) knife
10 Laundry marker
11 Road chalk
12 Road nails
13 3 m tape
14 Sledge-hammer
15 Theodolite
16 Level
17 Staff
18 Pegs, profiles and stakes
19 Ranging rods

20 Targets
21 Stationery
22 Clothing
23 Optical site square ('popeye')
24 Cigarettes/matches/confectionery
25 Abney level
26 Calculator

Of this list, items 1–13 should be carried in a bucket, preferably rubber. Items 14–20 are carried by hand (or in a vehicle) and 21–26 on the person of the site engineer.

Torpedo level

A good-quality torpedo level approximately 250 mm long, with three vials (0/45/ 90°) should be used. Necessary for placing stakes, pegs and profile boards correctly and squarely, it can also be used in quality control checks on construction work in progress.

30 m tape

Probably the most important piece of equipment, this item deserves the utmost care. Used in almost every conceivable situation, a modern glass-fibre-reinforced plastic tape will give a very high degree of accuracy and will survive great rigours. In some situations a steel tape, graduated in millimetres, is necessary to achieve a little more accuracy and one should be available for these special uses (and a refill or repair kit), but should not be in everyday use. A glass-fibre tape is usually graduated at 10 mm intervals (see Fig. 1.1) and the intermediate 1–9 mm points must be estimated, but this can be achieved, with considerable accuracy, with practice. A particular advantage of this type of tape is that it can be marked easily by the engineer with ball-point pen or pencil, making it unnecessary to rely upon an inexperienced chainman's ability to read a tape. These tapes usually have a folding clip at the end – very useful for measuring from walls, etc. when alone. However with the clip folded flat and placed over a nail, it should not be forgotten that an allowance for the clip (usually 5 mm) must be made.

50 m band

This item of equipment (see Fig. 1.2) is used less often, but is still very important – consisting of a 50 m steel tape graduated in millimetres, it is mounted in a special frame incorporating a handle, and has a rewinding mechanism which is often arranged to lock the band. The 'zero' end of the tape normally has a simple 'eye' or loop, which is intended to accept a spring balance to tension the band accurately. Used mainly for setting-out main control points as accurately as possible, it has the additional advantage (over the 30 m tape) of extra length, but

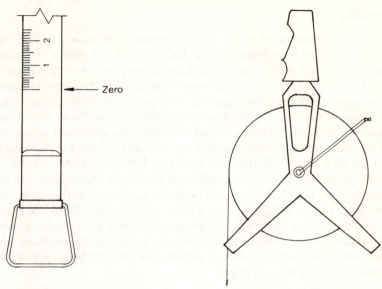

Fig. 1.1 30 m tape   *Fig. 1.2* 50 m band

is easily damaged. Proper instruction should be given to chainmen before use as the 'zero' mark is not at the extremity of the band, but usually about 100 mm in. (This can result in very large errors if not fully understood.)

Claw-hammer

A good-quality carpenter's claw-hammer is needed for construction of proper profiles and travellers.

Junior hacksaw

Although only occasionally necessary a saw is sometimes required, and if such a tool is not available immediately, time and money can be wasted. This little tool, consisting of a chromium-plated mild steel frame and a short but effective blade, will cut almost anything, and is therefore a great asset.

Nails

Nails are used for many purposes in site engineering, and it is better if only one size and type can be used for all of them – generally 50 mm roundheads (3 mm dia., 11 SWG) are adequate for profile construction, but also slender enough for 'pinpoint' requirements. They should be purchased by the box (25 kg), kept in the site office, and a handful thrown into the bucket when required.

3

Level-book

Level-book(s) should be of ready-printed pattern (collimation or rise and fall, depending on the engineer's preference) hard-backed and weather-resistant. Entries should be in ball-point pen or pencil, but in practice, pencil entries are better since they do not smudge when the book is closed, and they can be made in the rain. The book should be clearly marked on the *outside* with its sequential number (where a contract is likely to need several books) and the name of the engineer. Inside, on the front facing page, should be written the address to which the book should be sent and a telephone number, in the event of loss. Level-books tend to be used for all types of information (including levels), and this is excellent as a permanent record of important details is produced, in date order. Level-books occasionally form important pieces of evidence and should therefore be carefully filed and stored when full. Never more than one level-book should be kept in use at any one time – if a book is lost, or commandeered as evidence, a new book should be started (next sequential number). Do not restart the old book when it is returned; write the reason for discontinued use on the next blank page and deface the remainder of the unused pages.

200 m line

This is bricklayers' line, purchased in skeins which are invariably unravelled with great difficulty. (Read label instructions carefully before unravelling: there is one right way and several wrong ones.) Used for many purposes, this item is highly prized by subcontractors. Do not try to unravel the skeins immediately before use, much valuable time is wasted this way – the line should be prepared in a spare moment, and wound on to a bobbin for later use.

Arrows

Made of steel wire with a loop at the top and brightly coloured, arrows were originally used in 'chaining', to hold one end of the surveyor's chain to the ground. They can still be used for this purpose with a glass-fibre tape (making a suitable allowance for the clip), and are also useful in setting out profile/stake peg positions rapidly, to which the chainman can return later to place pegs and stakes. Usually sold in sets of twelve, they are rapidly lost – meat skewers are cheaper and just as good if brightly painted. Arrows are very useful as 'pointers' where a rusty or indistinct target nail has to be sighted from a distance.

Retracting (Stanley) knife

Useful to the site engineer for pencil-sharpening, string-cutting, etc. and whilst 'sheath' or 'Bowie' knives are pleasing items, a Stanley knife is probably more practical in site engineering.

Laundry marker

A vital piece of equipment – used to mark information on timber profiles and pegs to be later read and understood by foremen/gangers, etc. Generally having only a short life in this type of environment, laundry markers and other large felt-tip pens can be revitalised by filling with indian ink, but all inks should be waterproof.

Road chalk

This is a waxy compound (usually yellow, but other colours are available) which withstands rainfall and is used to mark concrete, tarmac, steel, etc. – usually used to highlight something which is to be relocated later.

Road nails

These are thick-bodied, chisel-pointed nails with large heads, useful for pinpointing main or subsidiary control points in roads, paved areas or brickwork. Pipe nails (used in fixing cast-iron drainage pipe brackets) are a good substitute.

3 m tape

An ordinary retracting tape-measure, used in a range of applications, is necessary for situations where the 30 m glass-fibre tape would be too cumbersome or too inaccurate.

Sledge-hammer

A '7 lb' 3.2 kg hammer is suggested for normal use, modified by cutting the handle shorter, so that pegs can be driven into the ground one-handed, the other hand holding the peg. Shortened, this hammer can also be used fairly easily for driving nails, and this can save considerable time and effort, saving the necessity for a claw-hammer to be carried around.

Theodolite

Basically, there is no outstanding difference between most modern makes of microptic theodolite in terms of their usefulness to the site engineer. Provided that the instrument is in good order and correctly adjusted, any modern theodolite will be capable of far more than a site engineer should ever be asked to do with it. There is, of course, a place for very precise equipment in other applications, but twenty seconds of direct reading is quite sufficient for all normal building/ civil construction contracts, and in most cases, direct reading of only one minute is enough.

The three most important factors affecting the site engineer's accurate use of a theodolite are:

1. The instrument must be properly adjusted, dry and clean
2. The site engineer must be familiar with the instrument and know how to use it in theory as well as practice
3. The instrument should be complete with plumb-bob and legs in full and proper working condition

Level

Three basic types of level are useful to the site engineer:

1. Three-screw
2. Quickset
3. Automatic

The three-screw level is set up in much the same way as a theodolite and whilst becoming obsolete, is useful for long-term use on the same spot, since it requires no intermediate adjustment once set.

The quickset level, probably the most versatile of the three, operates on a simple 'ball-and-socket' arrangement, and requires collimation corrections every time a reading is taken on a new horizontal angle.

Automatic levels are set up (usually) on three screwed feet, but only an 'approximate' centering of the spirit-level is required – the level does the rest. One common problem with auto-levels is that when the wind is blowing, the automatic levelling apparatus tries to compensate for every little buffet and nothing can be seen through the telescope.

As with theodolites, almost any make or type of level will suffice, if properly functioning – provided that the instrument is clean and dry, and the engineer is familiar with it.

There are other types of level available, but they are either for specialist purposes only, or not in common use. The precise level, which allows for readings of fractions of a millimetre in conjunction with a special staff, is (for instance) used by the Ordnance Survey for levelling bench marks, and can be used commercially for precise station levels. Laser levels of the datum type are sometimes used to provide a 'flashing' artificial datum on construction projects, but are adjusted conventionally.

Staff

A staff can be used with both level and theodolite, and is now normally available in both steel and aluminium, often folding neatly into a car boot (1 m) package. A surveying staff is merely a large ruler, but it must be read from a distance and therefore its markings are important; the familiar BS 'comb' or 'E' pattern staff as shown in Fig. 1.3 is widely used, and is a very effective system. There are

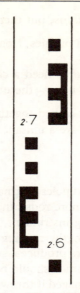

Fig. 1.3 Levelling staff markings

variants, but it is important only that the site engineer be familiar with the markings of his own staff. Graduations are usually in centimetre intervals, requiring estimation of millimetres, but this can be achieved with a high degree of accuracy with practice. It is essential that it should be capable of collapsing to a suitable size for transport.

Pegs, profiles and stakes

Construction management tend to view all peg-use as waste, and will occasionally emphatically question site engineers' requisitions for them. These items are generally held to be rather unimportant 'bits of wood', but they are in reality the only contact between the designer's efforts and the digger bucket. Besides being surprisingly expensive and remarkably short-lived (maximum life of a peg is two uses) they are also vulnerable to abuse as foot-scrapers, coat-hangers and perches – but are still expected to fulfil their original function with the utmost precision when their moment arrives.

To minimise the abuse it is helpful to paint them in bright colours (long before they are taken out on site) – so that they appear to be important pieces of equipment. Typical sections:

Pegs   50 × 50 × 600 mm Pointed
Stakes   50 × 50 × 1200 mm Pointed
Boards   75 × 12 × 900 mm

– these sizes cater for most situations, but there are exceptions:

(a) Sub-base and duct excavation profiles, (roads) normally require boards only 450 mm long
(b) Some complex foundation works need a continuous peripheral profile supporting hundreds of nails and notes – these are generally in 1800 mm lengths

Pegs, stakes and boards are best lashed together with an old belt or held in a canvas kitbag — loose pegs, etc. are a nuisance.

Ranging rods

Ranging rods are brightly coloured poles, normally used for marking or 'pointing' distant control points. Three or more rods can be used in straight-line 'ranging' with or without a site engineering instrument.

Supplied generally 2 m in length, ranging rods are cumbersome items to transport, but some manufacturers produce a screw-together rod in 1 m sections which will go in a car boot. Chainmen should be instructed that ranging rods are not javelins and will in fact be damaged if thrown – a very dangerous practice on a crowded construction site.

Targets

Where ranging rods may be used successfully in most situations, some applications (e.g. main horizontal control, traverses, etc.) demand the use of an aligned and levelled 'target' on its own set of legs. The equipment is aligned vertically over a ground-level station, using a plumb-bob.

Stationery

Bear in mind that the site engineer must communicate in diagrams, figures and words with equal facility, often in inhospitable circumstances. Only a few simple items are required to achieve a very high level of communication:

(a) Pencils (HB, H or 2H)
(b) A4 blank white paper
(c) Ball-point pen
(d) Clipboard

The clipboard should have clips to hold pens/pencils, and should be supplied with a waterproof cover to protect the paper, and a large rubber band to hold the lower edge of the paper to the board in a breeze.

Clothing

Faced with the vagaries of the British climate, correct choice of clothing for the site engineer and his chainman is vitally important – unlike the various tradesmen and gangers, the engineer finds himself rooted to one spot often for an hour or more, peering through a telescope, performing arithmetic or simply thinking. In cold weather it is important to insulate the body well, and often two pairs of trousers and two shirts can provide a sound foundation. Jeans are often favoured for their hardwearing qualities, but they do not insulate well and take a long time to dry out if they become wet. A combination of man-made fibre and wool is better, and usually lighter in weight than the cotton denim fabrics.

Boots and shoes, again a matter of taste, should be selected with the terrain in mind. Wellington boots (usually issued on site) are unsuitable for walking long distances, and it is often better to use a good-quality pair of lightweight leather upper/rubber sole boots for all situations except the most extreme, where full 'waders' would be more appropriate.

Hats are something of an affectation with some site engineers, who indulge in sombreros, fezzes or bowlers – others are satisfied with a more mundane ski-hat in wool, or a neat cap. A good deal of body heat is lost through the head, and a hat of some kind is essential in poor weather.

Outer garments must be waterproof and windproof. Plastic garments are tempting in that they appear to satisfy both conditions and are relatively cheap, but they do not 'breathe' and can become very uncomfortable in use. Garments of the 'overcoat' style are generally unsuccessful, as mobility is restricted. The Armed Services have been faced with a very similar set of personal clothing requirements for 'in-the-field' purposes, and a range of well-designed high-quality ex-service clothing can be found in shops specialising in these items.

Optical site square ('popeye')

This is a highly versatile, hand-held instrument comprising three prisms set in a moulded plastic body – the whole thing is small enough to be held in the palm of the hand. With this instrument it is possible to set out 90° angles from any given straight line with a surprising degree of accuracy. Some engineers do not like this uncomplicated little instrument, while others use it at every opportunity. The principle is very simple – the three prisms are arranged vertically in a stack. The top prism looks left, the middle one straight ahead and the bottom one looks right. In Fig. 1.4, by 'setting up' over point B on a line A–C (with ranging rods at A and C) the 'popeye' is moved back and forward until the images of rods A and C appear to intersect. Point D may then be set up by moving a ranging rod in the direction of D until all three images coincide. The angles ABD and CBD are then 90°.

This instrument is particularly useful for setting out profiles in road works from centre-lines and will work just as effectively from curved centre-lines, provided that AB–BC distances are the same.

The instrument is usually supplied with a long cord, and in a separate zipped pouch. The cord can be looped on to the pouch zip, and when in use, conveniently worn around the neck.

9

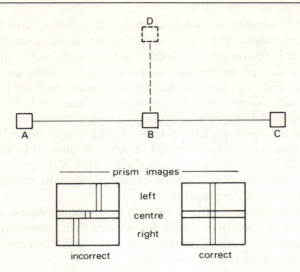

Fig. 1.4 Setting out right angles with optical site square

Cigarettes/matches/confectionery

Some engineers are lost without a cigarette, or something to chew. Checking in these items as 'equipment' removes the possibility of time-wasting or poor concentration.

Abney level

A versatile hand-held instrument, the Abney level comprises a spirit-level, and protractor with vernier facility. Vertical circle readings can be taken in seconds, where to set up a theodolite would normally take several minutes.

This instrument can be used to align sloping batter rails at cut/fill areas, or can provide the vertical angle for calculation of the height of a building, tree or spoil heap, with quite reasonable accuracy.

Calculator

Whilst the site office may be equipped with computer or printing calculator, the site engineer should have his own pocket machine for use outside. LCD (liquid crystal diode) display calculators can be seen in direct sunlight and have very low power consumption, and a full scientific format is highly desirable as this enables angular calculation to be carried out accurately on site. Calculators are now very cheap and extremely reliable. As there are many makes and models available it is not possible to be too specific, but a few points might be helpful in making a suitable choice.

1. Some types have a radian–degree-grad 'mode' button, which electronically switches in a conversion circuit to express your result in the desired format. With this type it is necessary to reset each time the calculator is switched on. Others have a sliding 'mode' switch, which does the same job as before, but presets the correct mode without resetting, and this is preferable as possible errors are avoided.

2. Ensure that the DMS (degree, minutes, seconds) system is simple to use – some machines will not accept DMS *entries* so these must be converted to decimal degrees before use. Others produce confusing outputs which, if not easily understood, can lead to errors.

3. To test the potential 'losses' caused by approximations and truncations in the chip, enter number 10, press sin cos tan and then function $\tan^{-1}$, function $\cos^{-1}$, function $\sin^{-1}$ – the closer the final figure is to 10 the better, but if more than (say) 0.4 per cent adrift in either direction this is unacceptable for accurate work.

Other equipment and accessories

Vehicles

Since he must be mobile, particularly on a large or fragmented site, a vehicle must be available for sole use by the site engineer and his chainman. It may be that a wheelbarrow will be sufficient, or at the other extreme a Range Rover (or even an aircraft) may be justified, but typically a lightweight commercial van will be supplied for normal site use, and this will usually be quite adequate.

Whatever vehicle is employed it must be made clear that this is for sole use of the site engineer, and not only will it contain essential equipment but it will be needed at unpredictable intervals throughout the day. Without a suitable vehicle at his disposal, the site engineer is unlikely to be able to respond adequately, if at all, to site requirements. For a large team of site engineers (e.g. on a motorway contract) it is usual to provide the senior site engineer with a 'crew bus' to ferry his engineers, chainmen and equipment from place to place as circumstances dictate, leaving him free to patrol and 'troubleshoot' between drops.

Radio telephones (RT)

Even on a very small project, RT can be a great time-saver, but on a large project it is usually essential. It can also be a useful asset for communication between engineer and chainman, particularly in noisy situations or when large distances are involved. Much time and effort can be wasted through incorrect interpretation of hand-signals, and shouting oneself hoarse at a chainman destroys concentration. The RT equipment is sometimes crossed off the list of 'essentials' by head office staff to effect 'savings' on the contract estimates, but it is usually possible to demonstrate (after the event) that the entire cost of the equipment for the duration of the contract would have been covered by savings on a single unpredictable site event (e.g. a calamitous concrete pour) where instant communication would have saved the day.

Computers

A site engineer's college or university course will probably have included tuition in both programming and operating. It is unlikely that the site engineer will be required (or even have the time) to design and write new programs, and it is more likely that any site-based computer will be supplied with a suite of 'library' programs. It is usually a mistake for the site engineer to spend too much time with the computer – use should be restricted to essential calculations only.

Drawings and prints

The main contract drawings must never leave the site office – this is an absolute rule. Ideally, contract drawings should be supplied in the form of secondary masters (negatives) and a plan-printer available for all prints to be issued for site use. If not available, an adequate supply of prints should be ordered from the client, and a new order placed whenever the stock appears low. These prints will normally be charged to the contractor, but this is a small price to pay for consistent availability of reliable information, and the alternative is intolerable. Secondary masters or main contract prints should be stored in darkness and hung from proper plan-strips – they will remain in the best possible condition this way, and the equipment needed can be purchased or made very cheaply. Under no circumstances should prints or secondary masters be stored rolled (except for posting) and *never* folded.

Chainman

The word 'chainman' is derived from the use of a surveyor's chain, which historically measures in links of 1 ft, the chain being 100 ft long in all, although there were variations on this. Now a chainman is usually a trainee engineer, often an undergraduate, and is a well-educated and skilled operative – it is fair to say that the best chainmen are in this category. Two qualified and experienced site engineers, one as chainman, working together, provide the perfect combination, as both understand precisely what is to be done and the reasons behind it, but this is rarely practical if only on grounds of cost. Inexperienced, unintelligent chainmen are rarely of use to the professional site engineer, and should not be considered. Bright, willing but relatively uneducated young people can often be useful after several weeks 'on the job' tuition, but the time involved in training a new chainman must be carefully considered before a commitment is made.

Care of equipment

Cleaning

Ideally, all equipment and instruments should be cleaned, dried and oiled (where appropriate) at the end of each working day. Under perfect conditions this is possible, but perfection is elusive. Under normal conditions a once-weekly general cleaning session is all that is possible, and this is usually enough, with the provision that instruments should be dried off whenever they become wet, and that steel tapes and bands should never be allowed to remain wet overnight. It takes about two minutes to wipe a theodolite dry, and about four minutes to clean a tape; a further two minutes is involved in running a steel band or tape through a hand-held oil-soaked rag. Optical instruments should never be oiled.

Instrument tripods are often made of wood and supplied thinly varnished. The varnish soon rubs off, the legs become mud-caked, dry, and warped, binding and useless. A weekly rub over with linseed oil will help to prevent this, and it takes only a minute or two. Do not be tempted to revarnish the tripod, as this can result in further binding of the telescopic sections and they may even become stuck together.

Instrument checks

Accuracy checks should be carried out as often as the site engineer thinks is reasonable – some engineers take the view that their level should be checked before each day's levelling, others check only when they suspect that something is wrong. Provided that 'closing' errors are insignificant, it can be assumed that the instrument is in adjustment, but if there is a large closing error or the instrument is in any way knocked or damaged, it should be checked immediately. Most contract firms expect their instruments to be sent for servicing and adjustment about once per year, and the instrument servicing firm will usually collect instruments and provide 'loan' instruments to replace them. Before using *any* new instrument, check it.

Level check procedure

Drive three pegs A, B, C into the ground in a straight line at 15 m intervals, as shown in Fig. 1.5. Set up at position 1 and take levels on A and C, record levels

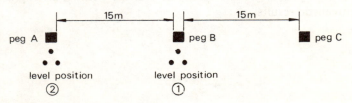

Fig. 1.5 Level check procedure

difference in level-book. Set up at position 2 and again take levels at A and C, record the difference. Any difference between the two results can be expressed as a collimation error over 30 m. The instrument may be adjusted by the site engineer if he knows how to do it, and the correct tools for this job are normally provided by the manufacturer, together with detailed instruction.

Theodolite check procedure

Vertical circle:

Carry out the procedure exactly as for a level, as shown in Fig. 1.5 – ensure that the vertical circle bubble is accurately centered each time

Horizontal circle:

(a) Set up the instrument, sight on to a distant (single) point, and read off the horizontal bearing

(b) Loosen the trunnion axis screw and rotate the telescope through 180° so that it points the other way

(c) Loosen the horizontal circle and turn through 180°, sight on to the original target point again and read off the new horizontal bearing

If the difference between the two readings is, say, 180°1'40", then the inaccuracy of the instrument is expressed as 50", being half the error shown by the readings. This is termed the 'face left–face right' procedure, and should be employed continuously in critical fieldwork, for example traverses or principal ground control.

The employer's responsibility

It is said that a poor workman will blame his tools, but in the site engineer's case the opportunity to blame equipment for errors should never arise. The engineer *must* be fully and properly equipped before he can begin to fulfil his function, and this is expensive in terms of both initial capital and maintenance. Particularly annoying is the occasional unexpected cost of repairs and replacements due to accidental damage or theft – these costs can be covered, for the most part, by insurance policies. A properly equipped and rewarded site engineer will save all of his initial cost, and will usually show the contractor a substantial profit. A poorly equipped and ill-rewarded engineer will show the contractor a very definite loss – it is probably better to try to manage without an engineer than to expect unrealistic results.

Chapter 2

In the beginning . . .

Setting up the site

Ideally, the engineer should allow several days on a new contract site prior to arrival of the main excavating equipment, during which time he can carry out some preparation work:

1. Co-ordinate 'control' points on a traverse round the site
2. Transfer of ordnance bench mark (OBM) levels to convenient temporary bench mark (TBM) locations around the site
3. Set out and supervise ground clearance and reduced-level dig to controlled tolerances, and arrange for 'muck-away' (spoil carted off site), 'site tip' (spoil to be retained on site for later re-use, e.g. topsoil or good backfill), and tip locations
4. Effect tree protections and disposals in association with local planning authority representatives
5. Make contact with building inspector, highways inspector, etc. and issue any statutory notices required for inspections, preliminary meetings on technical points as appropriate and raise any initial queries with the client's professional advisers (architect/structural engineer/services engineer, etc.)
6. Agree boundary and excavation base lines with client's representative and planning authority; record agreements by exchange of letters
7. Check all instruments, equipment and tools, paint adequate supplies of pegs, stakes and profiles and familiarise the chainman with all aspects of his duties
8. Check on locations, weights, depths and application of all services and make a note of all public utilities service personnel responsible for installations on the project site
9. Construct principal temporary access roads between site entrance and materials storage areas
10. In association with the site agent or general foreman, determine the correct location for brick stacks, concrete sections, steelwork, pipes, frames, etc. and plot these on to a master site plan (titled 'materials'). If done properly at the outset, double-handling and losses can be virtually eliminated

All too often, however, 'several days' are simply not available, and perhaps only a few items can be completed before the excavators arrive.

If you can keep your head . . .

At this point it is vital to keep cool, select what must be done first, then do it, immediately, with great care. Engineering priorities must be examined, assessed and decided *by the engineer*. On any site, and particularly on a completely new site, it is usually possible to find four hours' alternative productive work for any machine, operator or operative, to allow a little time for detailed preparation elsewhere – the engineer should not be afraid to use his authority to have works carried out in *his* preferred sequence, particularly where he needs a 'breathing space'.

Circumstances will dictate the priorities in items 1–10 above, but the most inflexible matters are the legislative ones (4, 5 and 6) and failure to attend to these can result in very serious contractual and legal problems later on. Where no 'free' time is available prior to contract, it will still be necessary to organise these items during the first week or two, probably outside normal site hours. It is well worth spending a few evening or weekend hours in this kind of activity, at the commencement of a contract, but it should be said that generally prolonged weekend and evening working should be avoided on the basis that a normal full working week, extended, physically and intellectually, is enough for any site engineer, and excessive hours will lead, inevitably, to dangerous or expensive mistakes.

Site clearance

This is usually the first operation involving an excavating machine, and it is important to ensure that both site engineer and operator are aware of the objective. It is unlikely that the entire site is intended to be stripped of all topsoil, plants and trees, but without clear directions the machine operator might decide to rip out everything and carry out a general 'reduced-level' dig for good measure.

On many sites there are existing trees and buildings to be protected, and operators must be told about them at the outset. These 'no-go' areas should be roped off with bunting and timber posts, the details marked on a plan (titled 'protected areas') and shown to all personnel involved.

Topsoil will usually be retained on site in a spoil heap, and great care should be taken in selection of a suitable location for it – a spot should be chosen which is not on the route of any main services, drains, roads or in an area to be built on. Transport of topsoil from the point of site strip to the spoil heap will probably be by tipper lorry, and the same lorry will no doubt be used to take 'rubbish' (tree roots, boulders, scrub, old bedsteads, etc.) to a local refuse tip off-site.

Security

It would be reasonable to hope that most of the 'rubbish' is disposed of at the off-site tip, and that most of the topsoil be left on site, but random inspections as the lorry leaves may disclose that a high proportion of the load is first-quality topsoil, and it may be suspected that a cash customer has been found to buy the stolen material. Appropriate action should therefore be taken, because at best the driver and/or machine operator are wasting precious project resources and storing up a problem for later (costly) solution. Check that there is no effective Statutory requirement in force for topsoil replacement.

Tipping charges

Tip charges can be a source of some confusion – it is often a recipe for anarchy to allow the lorry-driver to take money from the site office to pay in cash to a tip supervisor. Some form of lawlessness is almost inevitable in this situation.

A good method for dealing with payments for off-site tipping on a 'by-the-load' basis is to arrange with the tip manager to accept a voucher from the lorry-driver for each load tipped, on completion, the vouchers to be counted at both ends, agreed, and a cheque issued from head office – the vouchers to be numbered pages taken from a duplicate book issued by the site clerk for each load (one at a time), each one bearing the site stamp and signed. This method works just as well for materials being brought into the site, when paid for by the load, where even more money can be lost through uncontrolled inaccuracies.

'Unofficial' tips

It may be that the nearest 'official' tip is many miles away, and therefore rather uneconomic – a personal approach to the council offices can sometimes reveal an 'unofficial' (but fully legal) tipping site, and approaches to more than one department within the council offices should be made. A telephone call to all the farms in the vicinity can sometimes unearth a pond that needs filling or a low-lying area that needs raising – if the distance is less than the 'official' tip, then the excavation rate can be that much faster, and it is unlikely that a farmer would charge for tipping if he actually wanted the material. In any event, it is certainly worth spending thirty minutes with a telephone and the yellow pages.

Advertising

A week or two ahead of the commencement of clearing operations, advertisments could be placed in local newspapers offering free 'fill' materials – great care should be taken in accurately describing the quality of the spoil, and it should be made absolutely clear that this is not topsoil and must be taken in complete lorry-loads of 8 tonnes or more at a time. Access is also very important and telephone enquirers should be given full and clear information (even if they do not ask for it) so that confusion is minimised.

Reduced-level dig

For as long as possible, grassed and lightly overgrown areas should remain in their natural state – as soon as they are disturbed, their natural drainage and evaporation patterns will be destroyed, the soil beneath exposed and (depending on weather conditions) they will be either dusty or muddy patches of considerable nuisance value.

As the first part of the construction sequence, however, reduced-level dig is usually required over the whole area to be occupied by a building, and (as often applies) the simplest method is best. Figure 2.1 shows the outline of a future building, and ranging rods are used at the corners of an outer 'enclosing rectangle' giving an allowance all round the building for ease of access, scaffolding and to relieve the problem of excessively towering 'banks' immediately above excavated trenches.

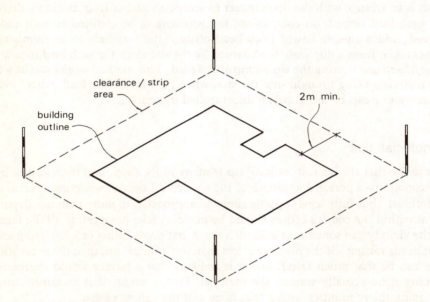

Fig. 2.1 Reduced-level/oversite strip

Formation Level: definition

Formation level is usually defined in site engineering as the top of the naturally occurring subsoil material, upon which new loads and imported materials are to be placed. It is important that formation levels be carefully calculated by the engineer. If too high, hand excavation may be necessary at a later date. If too low, expensive infill may be required.

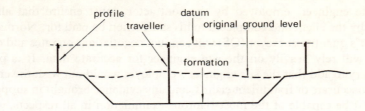

Fig. 2.2 Reduced-level profiles

Profiles and travellers

Figure 2.2 shows the usual arrangement of profiles for a small area of accurate reduced-level excavation. The two profile stakes are positioned and levelled, and profile boards are fixed on them at a convenient height from the ground – this is particularly important, because if not at a sensible height the operator will be reluctant to use them, and will simply guess at his finished level. The 'design' formation level is calculated as a distance below the horizontal datum line provided by the two profiles, and from this figure a 'traveller' of the correct height may be constructed. For single-handed operation (by the machine operator alone) a set of four legs may be added to the 'traveller' to allow it to stand upright on its own.

Level changes

It is sometimes desirable to change the oversite level from that quoted by the designer, particularly where the project is an isolated building, in order to achieve excavation/backfill economies, or for some other reason. Often the design level has been chosen from small-scale interpolated contour lines, and the actual 'existing' ground level on the site may differ considerably from that imagined at the design stage. Great care (*very* great care) should be taken before recommending new relative levels, being mindful of:

(a) Gas services new/existing
(b) Electricity services, underground and overhead new/existing
(c) Telephone services, underground and overhead
(d) Water services, pressure mains
(e) Foul drainage services and main sewer levels
(f) Surface-water drainage, sewer levels and groundwater (water-table) levels
(g) Associated roads, landscaping, drives and footpaths
(h) Steps, staircases and ramps

When agreed with the client's representative, any new oversite level should be confirmed in writing – it may be that a variation order (VO) will be required to allow for additional or reduced work to be carried out, and it is 'in the foundations' that a good proportion of construction profits are made or losses incurred.

The site engineer, employed by the contractor, must ensure that all work specified by the client and completed to his satisfaction be paid for. Normally the contractor's quantity surveyor (QS) will conduct negotiations on rates and quantities, but will rely heavily on the site engineer for accurate data. It is perhaps unnecessary to add that the site engineer's employer would not wish the engineer to make inaccurate or fraudulent claims, and any evidence brought in support of a claim must be capable of the most rigorous examination in all respects.

Figure 2.3a shows a typical 'design' construction detail and indicates that the 'design' formation level for this example would be at 250 mm below DPC. This is an idealised situation, and site conditions vary so much that it is very rare to have the finished product coincide exactly with the designer's typical cross-section.

Figure 2.3b shows an example of the 'design' detail adapted to meet site conditions – some additional peaty subsoil lay beneath the theoretical formation level as defined by the designer. The full extent of the unacceptable material should be explained and consideration given to reducing the 'design' DPC/ oversite level (bearing in mind the possible difficulties referred to earlier). If

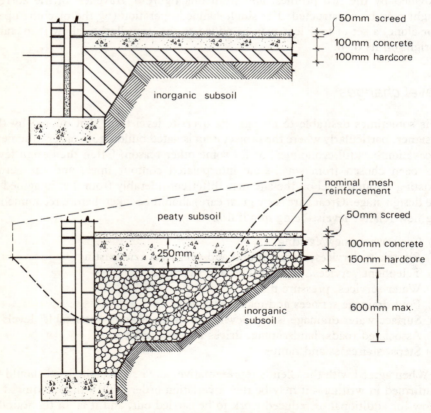

Fig. 2.3 Level changes: (a) design detail; (b) design adaptation

limited to a small area it would be reasonable to simply excavate and backfill with high-quality hardcore. It should be noted here that a backfill depth of greater than 600 mm will normally require the use of a suspended form of oversite level construction, or perhaps backfill in concrete. In the case of this example, additional hardcore in the affected area is quite acceptable and, subject to the designer's agreement, the remainder of the backfilling can be reduced in thickness from 150 to 100 mm owing to the clay-bound gravel (found under the peat) having a high enough bearing capacity to satisfy the building inspector.

Careful notes should be kept of these alterations to the designer's specification, together with memoranda of discussions held and agreements reached with all parties involved.

Trees

Trees can be a fascinating study and there are many trades and professions which deal exclusively with aspects of arboriculture. One of the principal efforts of the UK Forestry Commission is continually to replenish and upgrade stocks of standing timber 'sufficient to meet the country's needs in the event of a National Emergency' – to paraphrase a section of its charter. Many organisations, public and private, cherish trees as part of the 'national heritage' – some see them as protection for wildlife, as suppliers of oxygen, as producers of food or even as a source of veneers for domestic furniture. Large numbers of trees are maintained simply for their pleasing appearance.

Tree preservation orders

The UK Town Planning Acts allow for preservation orders on trees, as individual specimens, clearly defined groups or whole areas (referred to as a 'blanket' preservation order). Tree preservation orders have weight of law, and violations can, and often are, actioned by local authorities in court, often resulting in heavy fines. In setting up a conservation project it is best to involve both planning officer and tree preservation officer in a site visit to inspect each tree to be maintained or removed. Much work on this may have been done as part of the planning process, and the client's representatives should also be present. Trees to remain standing should be clearly marked with a white-painted ring at eye level, right round the trunk. Specimen trees, those of particularly fine appearance or unusual origin, should be protected by semi-permanent fencing such that there is absolutely no doubt that the enclosure is a prohibited area.

Technical problems

To the site engineer all trees are a physical obstruction, and also a potential nuisance in various ways:

21

1. They can take up excessive amounts of groundwater in times of relative drought, causing soil shrinkage, particularly in clays
2. Roots can grow into substructures and around drainage pipes, disrupting and dislodging essential structural components
3. Needles and leaves can block stormwater drains channels, causing minor flooding, and even whole branches can fall off, blocking roads and footpaths, causing damage to structures and injuries to people in the vicinity
4. In extreme cases, entire trees can fall on to building, causing terrible damage and even loss of life – in certain situations there might be a risk of an isolated tall tree being struck by lightning, or a large cluster of trees being consumed by fire

These points should be carefully considered by the council officers and client's representatives at the site engineer's initial meeting on the subject, a suitable balance struck and the various components confirmed in a letter, preferably with a clearly marked site plan titled 'trees'.

Growth patterns

The specialist council officers should be able to advise on tree development and likely future shapes and sizes, but as a rough guide, the roots of a tree can be expected to extend to around twice the spread of its foliage at maturity, therefore a pine with branches radiating 2.5 m from the trunk at maturity might be sited at a minimum of 5 m from a structure, with reasonable safety. If in doubt seek specialist advice, or consider some exploratory excavation.

Tree surgery

It is possible to adjust tree development to suit site conditions, cutting back roots and branches periodically to 'train' a tree to suit its altered environment. This work should be contracted to a firm of 'tree surgeons' who should best be nominated by the local council – the processes are quite straightforward, but a great deal of skill is necessary in judging the proper time for the operation and its degree of severity.

Moving trees

It is possible to move even very large trees for distances of many miles and replant them successfully. Such an operation is very expensive, commensurate with the extent of the work involved, and is a task for specialists – it is, however, a feasible proposition under the right circumstances if no alternative is possible.

Advantages of trees

The site engineer may find that, despite their apparent drawbacks, trees may have advantages.

Conservationists may sometimes be satisfied with a well-considered scheme for new tree plantings, to replace existing trees to be removed. In these situations the conservationists usually do rather well, as they tend to achieve a larger number of trees than before, better sited and of the most appropriate species.

Trees (new or existing) can 'soften the edges' of a new piece of construction work and do generally improve the finished appearance.

They can be used to disguise or conceal an otherwise unpromising structure — a prefabricated sewer-pump installation, an electricity substation transformer unit, a refuse baler, etc.

Relationships in the site management team

As a general rule the site engineer obtains his authority from the site manager, and may issue direct instructions on engineering matters to all personnel at every level. His capacity is a 'functional' one, and does not have the strength and limitations of the 'line-of-command' relationship enjoyed by the foremen.

It is often advantageous to initiate a full management team meeting to discuss relationships on site, early on in the contract – whilst this is of no immediate benefit to the engineer personally, it can ease tensions, remove uncertainties, doubts and suspicions, and encourage a genuine understanding. Almost always these discussions are of general benefit.

In project management terms, the site engineer is a member of a team of people of varying abilities, capacities and interests – this does not mean that any member of the team is more or less 'important' than any other, indeed, a wide range of aptitudes is necessary to achieve an effective production-management structure. The site engineer's principal function is to supply technical information to everyone who needs it, to achieve the result that the designer intended and the client expects. Other members of the team have different functions – the site clerk must be skilled in record-keeping, office management and analytical techniques; he will take pride and pleasure in producing impeccable data-sheets from chaotic piles of dishevelled scraps of paper – the groundworks foreman will have the skill to organise and motivate numbers of tough men in the execution of feats of physical labour and endurance, and so on. Each team member is reliant on many others.

Attitude to the engineer

Foremen and tradesmen generally have a good deal of respect for even relatively inexperienced site engineers (and trainees) and most will genuinely go out of their way to assist. Instructions and requests that are clear, polite, definite and timely are all that are required in return, together with an occasionally more detailed explanation or discussion on a particularly complex aspect. Sometimes of course it will be necessary to reprimand a man – even a general foreman – and at other times to give praise. For either to mean very much, the recipient must hold the engineer in respect, or the words will be empty, and for this reason it is usually politic to restrict contact to commercial and professional matters – it is likely to be difficult to rebuke a foreman for (for instance) ignoring perfectly adequate setting-out information if the previous evening was spent socialising with him.

If a mistake has been made, this should be immediately admitted, and the entire matter reported to the site agent as a matter of course, particularly if any extra cost is involved or delays incurred. Having once identified an error, acknowledged or allocated the 'blame' and then remedied the problem, the matter should then be disposed of calmly and discreetly without further recrimination. Generally, out of every five decisions, three will be exactly right, soundly based on prior knowledge and experience, one might not be too significant either way and the remaining one stands a fifty-fifty chance of being right or wrong – remembering this can help the site engineer or the man he is reprimanding, to keep a sense of proportion. There is little to be gained (and much to be lost) by adopting a paranoic approach that 'everything must always be perfect'.

Site management structure

In the 'family tree' shown in Fig. 2.4, the arrangement is idealised, but it serves to illustrate the site engineer's relationship to other members of the team.

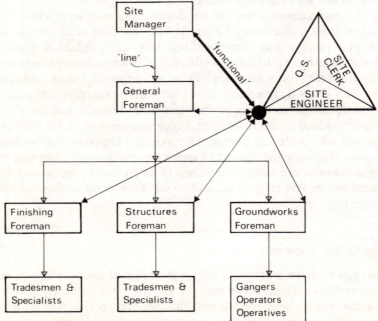

Fig. 2.4 Site management structure

With a site management structure of this size it would be possible to run a large mixed construction development, involving roads and sewers, housing and (for instance) small factory units, all at the same time. On a smaller site where for example, the production target is one house completion per week, the project may be run by a smaller team comprising agent, clerk and one or two foremen. On a site of this magnitude, engineering services would be provided on a 'visiting' basis (as would the quantity surveyor's services) and much of the routine setting-out and checking would be done by the agent.

The site engineer will not normally be directly concerned with site management structure arrangements, as his relationships are all 'functional' in principle (supplying technical services to all levels within the team) in contrast to the foremen's 'line' relationships where in theory there is a rigid chain of command. For this reason it is possible for a site engineer to slip in and out of several site management teams during the course of a working week, playing a constructive role in each, if this is what is required.

Responsibility and authority

A typical list, which is not a specific 'job description' but could provide a foundation for one, might be as follows:

Responsibilities

1. Provision of sufficient and adequate technical information on site to be used by foremen in directing and organising construction functions

2. Carrying out random measurement checks and other inspections to ensure that work is being carried out correctly, from the technical viewpoint, in co-operation with foremen

3. Maintaining excavation, progress and other site records for measured work for payment

4. Maintenance of tools, equipment and instruments used to fulfil all technical services requirements

Authority

1. Authorised to issue instructions on site engineering matters to all construction personnel

2. Authorised to sign time-sheets, commission overtime working, requisition materials and equipment and sanction expenses claims, all within broad limits and constraints previously agreed with the site agent.

This summary of authority and responsibility is brief, but should be sufficient to indicate the site engineer's approximate position in the site hierarchy. As with many professional roles, there has to be considerable goodwill on both sides for the relationship to be effective at all, but the individual site engineer should (to a greater or lesser extent) have the following qualities and abilities:

1. Open-mindedness, receptive to 'new' ideas
2. Restraint
3. Patience
4. Explanation/teaching ability
5. Calm/coolness
6. Courage of convictions
7. Logic, intelligence, deductiveness
8. Organisation and order
9. Authority projection
10. Command of respect from subordinates and superiors
11. Decisiveness
12. Leadership

Chapter 3

Structural calculations

The site engineer will occasionally be required to specify structural components, usually under pressure with little notice. In these situations it is preferable to 'over-design', since at this stage in the works there is no time for the lengthy deliberations possible in the consultant's design office, and no place for elegant theoretical proofs – either the component will do its job or it will fail, and as the site engineer deals only with problems at this basic level, there is a need for simplified design data and rudimentary worked examples leading to sound, if unrefined, solutions.

The role of the site engineer in performing structural calculations

There is no intention that the part of the site engineer should usurp the role of the structural design consultant – indeed, immediate reference should be made to the consultant whenever a problem has arisen with the main contract information – but there are many instances where the site engineer can resolve a minor problem within a few minutes, where a solution from the consultants might take a day or two to materialise. Any workings, calculations and assumptions should be carefully written down, and if applicable, a copy should be sent on to the consultants for verification and incorporation in the contract drawings – this is rather important if a claim for payment is to be made, and notification should be given in the usual way with a copy of all calculations, notes, etc. as evidence.

Negligence

The site engineer would, of course, be held liable for any negligent work performed – if employed by a contractor he would run the risk of losing his job, and a court action might arise.

A self-employed site engineer might carry a professional indemnity insurance policy, but would probably find that premiums and policy wording would be considerably altered at the renewal date following a claim.

It is essential for the site engineer to realise that when performing structural calculations he is temporarily assuming the mantle of a consulting engineer, and should enter into this unfamiliar arena with the greatest care. He must also be prepared for the consequences of his actions, as expensive machinery and valuable buildings, to say nothing of peoples' lives, are very definitely at risk. By all means achieve a reputation for generous but practical over-design – absolutely no one will be impressed by a structural failure.

Simplified treatment

It is not intended that this section should pretend to be a treatise on the principles of structural engineering – there are many fine books dealing exclusively with the theory and design of structures, and the site engineer should refer to one or more of these for a more extensive treatment on any one item of special interest. This chapter is concerned with the very basic problem of designing components to fulfil immediate needs on site, as and when the requirement arises. To this end, there has been drastic abbreviation, compression and omission of all but the most essential data; it has been assumed that (for instance) all members will be used in the 'upright' (XX) attitude, that all timber will be of GS or SS grades in species group 2 (S2), redwood/whitewood, etc.

Normal stock steelwork

The site engineer, by the very nature of his calling, will want his steelwork delivered to site and installed in position as soon as possible – very often on the same day that the original problem arises, necessitating his intervention.

There is little point in specifying steel sections that are only available on seven to ten days' delivery, or are only occasionally (or never) rolled at the steel mills. The usual steel tables give details of a wide range of sections, and with sufficient time for design work and procurement, consultants and contractors can achieve major economies through precise specification of the best possible components for specific tasks. However, a site engineer has little time for the niceties of the pre-contract design office, and it is therefore convenient to refer only to sections held as basic stock by most local steelwork suppliers and distributors. Table 3.1 gives tabulated details of sections commonly stocked, with abbreviated 'properties' data, commensurate with the limited requirements of the site engineer. Rather than designing 'forwards' to achieve the most elegant solution, the site engineer usually designs 'backwards' to make use of what is available.

Worked examples

The worked examples require constant reference to tables which are found on pp. 45–49 inclusive.

Example 1 Steel beam to carry a point load

A maximum point load of 3.5 kN, gently applied: span of beam 2.6 m between support centres, loaded centrally.

Bending moment (max) (from Fig. 3.2) $= \dfrac{WL}{4} = \dfrac{3.5 \times 2.6}{4} = 2.275$ kN.m

Z (elastic modulus) required $= \dfrac{2.275 \times 10^3}{165} = 13.788$ cm^3

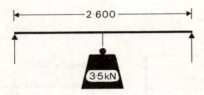

Fig. 3.1 Beam to carry point load

Check a section from tables (see Table 3.1): 102×64 (9.6 kg) has a Z actual of 42.84 cm^4.

The Pbc is now checked. From Table 3.1 the D/T ratio for this section is 15.4 and ry is 1.43.

The slenderness ratio $\quad \dfrac{1}{ry} = \dfrac{2.6 \times 10^2}{1.43} = 181.818$

The *allowable stress* table from BS 449 Part 2 (Table 3.2) gives a Pbc of around 124 N/mm^2 (180 on the l/ry and 15 on the D/T). The table values should now be interpolated in both directions to find a precise Pbc, and a figure of 120.669 is obtained. 120 N/mm^2 is used in this example.

Rerunning the Z required calculation using the reduced Pbc value:

Z required $= \dfrac{2.275 \times 10^3}{120} = 18.958$ cm^3

which is still much less than the 42.84 cm^3 offered by the 102×64 mm section, and is therefore acceptable.

The next stage is a consideration of deflection:

Maximum deflection (Da) $= \dfrac{1}{48} \cdot \dfrac{WL^3}{EI}$ (from Fig. 3.2)

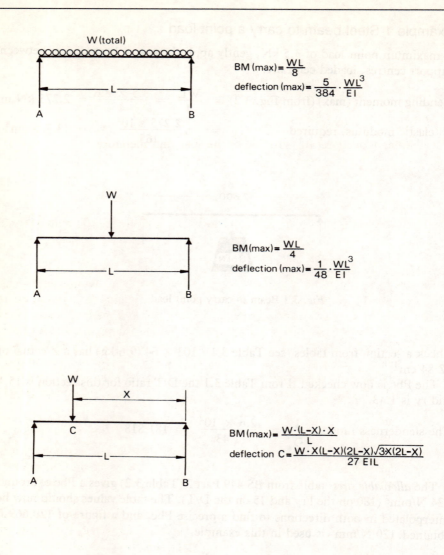

$$BM(max) = \frac{WL}{8}$$

$$deflection\ (max) = \frac{5}{384} \cdot \frac{WL^3}{EI}$$

$$BM(max) = \frac{WL}{4}$$

$$deflection\ (max) = \frac{1}{48} \cdot \frac{WL^3}{EI}$$

$$BM(max) = \frac{W \cdot (L-X) \cdot X}{L}$$

$$deflection\ C = \frac{W \cdot X(L-X)(2L-X)\sqrt{3X(2L-X)}}{27\ EIL}$$

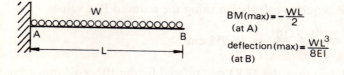

$$BM(max) = -\frac{WL}{2}$$
(at A)

$$deflection\ (max) = \frac{WL^3}{8EI}$$
(at B)

Fig. 3.2 Standard bending moment and deflection cases

Although this is only a prediction, it is referred to as 'actual' (Da) because it will finally be compared with a 'permissible' deflection figure, expressed as a fraction of the span.

$$Da = \frac{1 \times 3.5 \times 10^3 \times (2.6 \times 10^3)^3}{48 \times 2.1 \times 10^5 \times 217.6 \times 10^4}$$

$$= 2.804 \text{ mm}$$

Permissible deflection (Dp) is 1/360 of the span and therefore

$$Dp = \frac{2600}{360} = 7.22 \text{ mm}$$

Evidently there is no problem as Dp exceeds Da, and the section may be used.

Further considerations for Example 1
Attention should be given to suitable bearings on site which should accept point loads of around 2 kN including beam self-weight at each end of the beam, on an area of 6400 mm^2, so a bearing material capable of taking a working load of 0.31 N/mm^2 would be required. 1:2:4 concrete padstones offering a maximum permissible stress of 7.86 N/mm^2 should be built into the brickwork and the RSJ built in, bearing directly on to them. In an emergency the crushing strengths of other materials may be examined (e.g. facing and common bricks, concrete blocks, concrete lintels) as these can all make acceptable substitutes under certain conditions.

Lateral restraint Regardless of situation, it is advisable to consider some form of lateral restraint which will stop the beam moving sideways at the centre of its span, to reduce the risk of movement should there be any accidental rough handling or improper use during its working life. If it can be guaranteed that the beam will be laterally restrained, limiting sideways-acting forces to one twentieth of the downward-acting forces, then the Pbc figure can be taken as 165 N/mm^2 in every case, without the need for a stress-reduction exercise as seen in this example.

Maximum load for this case Sometimes it is necessary (or simply gratifying) to quote the maximum loading that can be placed on any particular beam – taking the above example we can arrive at a *maximum* load figure as follows.
    Inverting parts of the earlier deflection equation and substituting the maximum deflection value for the Dp figure, leaving W as the unknown:

$$7.22 \text{ mm} = \frac{W \times 10^3 \times (2.6 \times 10^3)^3}{48 \times 2.1 \times 10^5 \times 217.6 \times 10^4}$$

31

isolating W:

$$W = \frac{7.22 \times 48 \times 2.1 \times 10^5 \times 217.6 \times 10^4}{10^3 \times (2.6 \times 10^3)^3}$$

$$W = 9.010 \text{ kN}$$

The section modulus (Z) must now be checked.
From the equation:

$$\frac{2.275 \times 10^3}{120} = 18.958 \text{ cm}^3$$

(the 'Z required' calculation used earlier) the bending moment (2.275) is made the unknown, and the Z actual value 42.84 cm³ from Table 3.1 is substituted for the 18.958 cm³. This leaves:

$$\frac{\text{BM} \times 10^3}{120} = 42.84 \text{ cm}^3,$$

and this expression is rearranged to isolate the unknown:

$$\text{BM} = \frac{42.84 \times 120}{10^3}$$

$$\text{BM} = 5.14 \text{ kN.m}$$

going back a stage further to the original bending moment calculation:

$$\text{BM} = \frac{\text{WL}}{4} = \frac{3.5 \times 2.6}{4} = 2.275 \text{ kN.m}$$

W is the unknown, and substituting the maximum BM figure of 5.14 the equation is rearranged to isolate W:

$$(\text{BM}) \ 5.14 = \frac{\text{W} \times 2.6}{4}$$

$$W = \frac{5.14 \times 4}{2.6}$$

$$W = 7.908 \text{ kN}$$

There are now two 'maximum' values for W, 9.010 and 7.908 kN, and clearly the lower one will be reached first in a test-load situation, so 7.908 kN is taken. One final matter remains – an allowance should be deducted for the self-weight of the beam itself and any lifting tackle. The beam's own self-weight is around 0.25 kN and the lifting equipment probably about the same again. For argument's sake, although it could be more precisely calculated, a maximum load of 7 kN should be stipulated by the site engineer, gradually applied.

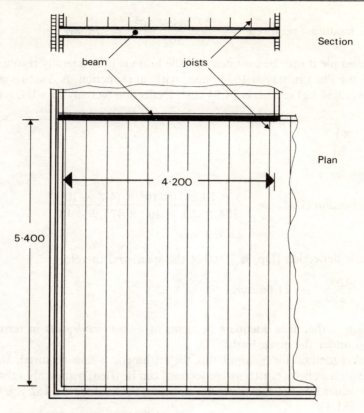

Section

beam   joists

Plan

4·200

5·400

Fig. 3.3 Beam supporting joists

Example 2 Steel beam supporting joists (see Fig. 3.3)

This is a common situation and can arise in many guises. Joists are assumed to carry a uniformly distributed load (UDL) to the beam.

Beam load is 2.05 kN/m² (Table 3.4) multiplied by the loaded area:

$$4.2 \times \frac{5.4}{2} \times 2.05 = 23.247 \text{ kN}$$

Bending moment (max) BM (max) $= \dfrac{WL}{8}$ (Fig. 3.2)

$$\frac{WL}{8} = \frac{23.247 \times 4.2}{8} = 12.205 \text{ kN.m}$$

Note: L should be the distance between support *centres* and not the clear span, but this has been ignored for this example.

33

Z (elastic modulus) required $= \dfrac{12.205 \times 10^3}{165} = 73.969 \text{ cm}^3$

for this example it may be assumed that the beam is fully laterally restrained and therefore the Pbc can stay at 165 N/mm² without reduction. A Z actual of (say) a 127 × 76 (13.36 kg) section is 74.94 cm³. Deflection (actual) for a UDL is based on

$$\dfrac{5}{384} \times \dfrac{WL^3}{EI}$$

and therefore:

$$\text{Actual deflection (Da)} = \dfrac{5 \times 23.247 \times 10^3 \times (4.2 \times 10^3)^3}{384 \times 2.1 \times 10^5 \times 475.9 \times 10^4}$$

$$= 22.439 \text{ mm}$$

Permissible deflection (Dp) is 1/360 of the span, and therefore

$$\text{Dp} = \dfrac{4200}{360} = 11.66 \text{ mm}$$

this section is therefore adequate in terms of Z, but *inadequate* in terms of its deflection under the given loads.

A quicker method for manipulating the arithmetic is now required. Using the same equation as before, alternative sections can be tried, using only substitution of their I values into the above result, i.e. (using the I value for a 152 × 89 (17.09) section – 881.1):

$$\text{New deflection} = \dfrac{22.439 \times 475.9}{881.1} = 12.119 \text{ mm}$$

This is still not enough, so a third section is tried, i.e. (using the I value for 178 × 102 (21.54) section – 1519):

$$\text{New deflection} = \dfrac{22.439 \times 475.9}{1519} = 7.03 \text{ mm}$$

this is a satisfactorily small deflection, compared with the 11.66 mm permissible, so the 178 × 102 (21.54) section may be used.

Further considerations for Example 2
The effect of the self-weight of the new section should now be considered, and it can be seen that in this case the beam's own weight adds approximately 90 kg or less than 1 kN to the applied loads, and it can be seen by inspection that the beam section chosen is still adequate with this additional loading considered.

Example 3 Timber or steel beam supporting an internal wall (Fig. 3.4)

This is a common design requirement in housing and office layouts. Generally, it is advisable to use a steel beam where a blockwork or brickwork partition is used, as timber sections tend to shrink and deform – slightly, but not always predictably. Whilst timber is an excellent structural building material for a wide range of situations, it is not always suitable in situations where long-term static bending stresses prevail.

It should be added that the phrase 'double joists under partitions where parallel to the span direction' is sometimes used on drawings – invariably this refers to lightweight studwork partitions, but if the detail is not absolutely clear the matter should be raised with the designer.

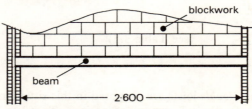

Fig. 3.4 Beam supporting wall

For the following example, a floor–ceiling height of 2.350 m is assumed. Alternative 'timber' and 'steel' solutions are considered to support a blockwork partition for the full span width (see Fig. 3.4).

The following details, from Table 3.4, are the same for both cases:

$$\text{Load} = 2.57 \text{ kN/m}^2 \times 2.6 \text{ m} \times 2.350 \text{ m} = 15.7 \text{ kN}$$

$$\text{BM} = \frac{WL}{8} = \frac{15.7 \times 2.6}{8} = 5.10 \text{ kN.m}$$

Note: L should be distance between support *centres*, but this is ignored in this example.

The timber case
(Assuming the use of GS, S2 redwood/whitewood)

$$Z \text{ required} = \frac{5.1 \times 10^6}{4.8 \text{ (from Table 3.7)}}$$

$$= 1062.5 \times 10^3 \text{ mm}^3$$

from Fig. 3.2, actual deflection (Da) $= \frac{5}{384} \times \frac{WL^3}{EI}$

everything is known except the value of 'I', and from a calculation of the *permissible* deflection Dp:

$$\text{Dp} = \frac{L}{325} = \frac{2600}{325} = 8 \text{ mm}$$

35

the value of the *required* I (I required) can be established:

$$
\begin{aligned}
\text{I required} \quad &= \quad \frac{5}{384} \times \frac{WL^3}{E \times Dp} \\
&= \quad \frac{5 \times 15.7 \times 10^3 \times (2.6 \times 10^3)^3}{384 \times 4300 \times 8} \\
&= \quad 104.448 \times 10^6 \text{ mm}^4
\end{aligned}
$$

A section must be selected which will satisfy both the 'Z required' figure of 850×10^3 mm^3 and the 'I required' value of 97.6×10^6 mm^4. The following formulae can be applied to various timber sections:

$$
\text{Z actual} \quad = \quad \frac{bd^2}{6}
$$

$$
\text{I actual} \quad = \quad \frac{bd^3}{12} \qquad \text{(see Table 3.5)}
$$

bearing in mind that *double* joists are proposed in this situation, only half of the total Z or I is taken for the calculation comparison. *Try* 63 × 225 mm section:

$$
\text{Z} \quad = \quad \frac{63 \times 225^2}{6} = 531.563 \times 10^3 \text{ mm}^3 \quad \text{(compares with } \frac{1062.5}{2} = 531.25)
$$

$$
\text{I} \quad = \quad \frac{63 \times 225^3}{12} = 59.8 \times 10^6 \text{ mm}^4 \quad \text{(compares with } \frac{104.448}{2} = 52.22)
$$

It is therefore shown that the use of two 63 × 225 mm GS (S2) joists will support the blockwork partition.

Further considerations for the timber case By reference to Table 3.5 in which properties of commonly used section are listed for rapid comparison, it can be seen that the values for (for instance) a 50 × 225 mm section are:

Z = 422×10^3 mm^3 (compares with 531.25 required)
I = 47.5×10^6 mm^4 (compares with 52.223 required)
. . . so this would be unacceptable.

The steel case
This calculation is very similar to that shown in Example 2:

$$
\text{Z required} \quad = \quad \frac{5.1 \times 10^3}{152} = 33.553 \text{ cm}^3
$$

(assuming a 178 × 102 mm (21.54) RSJ, l/ry = 115.56, D/T = 19.7 – allowable stress Pbc approximately 152) (see Tables 3.1 and 3.2)

Z actual (178 × 120 mm RSJ) = 170.9 cm^3

The inversion can now be used as in the timber case shown in this example:

$$Dp \quad = \quad \frac{L}{360} \text{ (steel)} \quad = \quad \frac{2600}{360} \quad = \quad 7.22 \text{ mm}$$

$$\text{I required} \quad = \quad \frac{5}{384} \times \frac{WL^3}{E \times Da}$$

$$= \quad \frac{5 \times 15.7 \times 10^3 \times (2.6 \times 10^3)^3}{384 \times 2.1 \times 10^5 \times 7.22}$$

$$= \quad 236.9747 \times 10^4 \text{ cm}^4$$

'I actual' for a 178×102 mm RSJ $= 1519 \times 10^4$ cm^4 (from Table 3.1), therefore adequate.

Further considerations for the steel case. From an inspection of the steel sections table it could be shown that (for instance) a 127×76 mm section might be adequate, but for convenience of construction it is preferable to supply a section at least as wide as the wall it supports, hence the recommendation for a 178×102 mm member in this case.

Example 4 Timber joists supporting a partition (at 90° to the span direction) (Fig. 3.5)

For this example two different loads are operating:
(a) A uniformly distributed floor load (ignoring any special loads from the bathroom fittings) for both bathroom and bedroom
(b) An eccentric 'point' load to each joist, the effect of the superimposed load from the studwork partition.

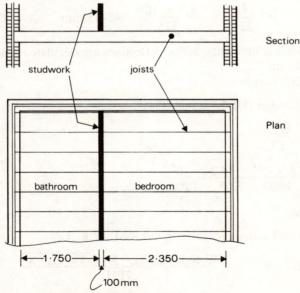

Fig. 3.5 Joists supporting studwork

Distances are taken to the centre-line of the partition in Fig. 3.5. For both loads, only one individual joist is designed. Load (a) (the UDL) is obtained from:

$$\frac{1.8}{2} + \frac{2.4}{2} \times 0.4 \text{ (400 mm centres)} \times 2.05 \text{ kN/m}^2 = 1.722 \text{ kN}$$

Note: L should be distance between support centres but this is ignored in this example.

Load (b) (the 'point' load) is obtained from:

$$0.4 \times 2.350 \text{ (room height)} \times 0.8 \text{ kN/m}^2 = 0.752 \text{ kN}$$

$$\text{(Fig 3.2) BM (UDL)} = \frac{WL}{8} = \frac{1.722 \times 4.2}{8} = 0.904 \text{ kN.m}$$

$$\text{(Fig 3.2) BM (point)} = \frac{W.(L-X).X}{L} = \frac{0.752 \times 1.8 \times 2.4}{4.2}$$

$$= 0.773 \text{ kN.m}$$

$$\text{BM (UDL)} + \text{BM (point)} = 0.904 + 0.773 = 1.677 \text{ kN.m}$$

$$\text{Z required} = \frac{1.677 \times 10^6}{4.8 \times 1.25} = 279.5 \times 10^3 \text{ mm}^3$$

(a factor of 1.25 is applied to reflect short-term live loading – 'mean' rather than 'minimum' value of E is taken when five or more sections are used together; value from Table 3.7).

Maximum permissible deflection:

$$\text{Dp (timber)} = \frac{L}{325} = \frac{4.2 \times 10^3}{325} = 13 \text{ mm}$$

as with the bending moments, actual deflections are calculated separately for each loading condition.

Try 50 × 200 mm joists:

(Fig. 3.2) deflection (UDL) actual (Da)

$$= \frac{5}{384} \times \frac{WL^3}{EI}$$

$$= \frac{5 \times 1.722 \times 10^3 \times (4.2 \times 10^3)^3}{384 \times 8000 \times 33.3 \times 10^6}$$

$$= 6.235 \text{ mm}$$

deflection (point) actual (eccentric point load) (Fig. 3.2)

$$= \text{Da (point, E/c)} = \frac{W.X(L-X)(2L-X)\sqrt{3 \times (2L-X)}}{27EIL}$$

. . . in this case X = 1.8 m, L = 4.2 m (see Fig. 3.2)

$$= \frac{0.752 \times 10^3 \times 2.4 \times 10^3 \times 1.8 \times 10^3 \times 6 \times 10^3 \times 6.5726 \times 10^3}{27 \times 8000 \times 33.3 \times 10^6 \times 4.2 \times 10^3}$$

$$= 4.241 \text{ mm}$$

Whilst the deflection (UDL) figure is calculated for the *centre* of span, the value obtained for deflection (point) using the above formula is not, but for the site engineer's purposes this will make no appreciable difference, and treating the deflection value as if acting at the *centre* will produce a result within 2½ per cent of the maximum. In fact, maximum deflection always occurs within 0.0774 L from the span centre of the beam towards the load.

Hence, aggregate deflection (actual) taken as:

6.23 (UDL) + 4.241 (point) = 10.471 mm

. . . against a maximum permissible deflection of 13 mm. Therefore: *Use 50 × 200 mm GS Joists at 400 mm Centres.*

Example 5 Timber and steel columns

The site engineer should be able to establish the effective loading capacity of a given steel or timber column, and this example provides calculations for both cases. A column of length 3.000 m is considered, axially loaded with 350 kN, 'pin-jointed' top and bottom.

Effective length
For site engineering purposes, the 'effective length' of a column should be taken

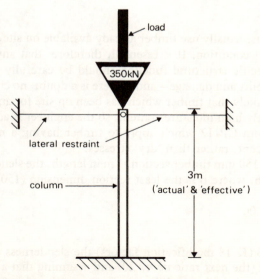

Fig. 3.6 Column supporting axial load

39

simply as 'l', the actual length – assuming that both ends are fixed in some way. When a column is unrestrained at the top, its 'effective length' should be taken as 2 × l. Other variations are given in the codes, but as the site engineer is not very concerned with 'economic' design, but needs a reliable and safe answer quickly, there will be a tendency to over-specify and such detailed considerations become irrelevant.

The steel case
Taking a 152 × 152 (30) universal column as an example, the ry is given as 3.82 cm in Table 3.1. The slenderness ratio is therefore:

$$\frac{l}{ry} = \frac{3 \times 10^3}{3.82 \times 10} = 78.534 \text{ (mm)}$$

from Table 3.3 the allowable stress (Pc) is 106 N/mm² (70 is found in the column on the left of the table, 8 is found at the top: 70 + 8 = 78).

The maximum load which can be taken by a column is the allowable stress (Pc) × the area of the column section as follows.

For a 152 × 152 (30) universal column, the maximum load would be:

$$W = \frac{38.2 \times 10^2 \times 106}{10^3} \quad (38.2 \text{ cm}^2 \text{ is the area of section from Table 3.1})$$

$$= 404.92 \text{ kN}$$

which exceeds the 350 kN requirement considerably, and the section may therefore be used. It should be noted that universal beams, rolled steel joists, etc. can also be used as columns.

The timber case
A site engineer must usually use timber already available on site, and it is rare to find this in perfect condition. It is essential, therefore, that any timber section intended for a specific structural function should be carefully inspected on all faces for cracks, splits and damage – and if there is a doubt, no chances should be taken. It is quite likely that timber which has been on site for any length of time will have a relatively high moisture content, so the engineer is advised to use the 'green' stresses from CP112 which apply to timber having a moisture content exceeding 18 per cent, rather than 'dry' stresses.

Taking a 425 × 150 mm timber section, 3 m in length, the slenderness ratio can be expressed as l/b, where b is the least section dimension (150):

$$\frac{3 \times 10^3}{150} = 20.00$$

From Table 3.6 (K 18 modification factors) the slenderness ratio 20.2 is the nearest equivalent (the next ratio *above* 20) and assuming that a 'long-term load' arrangement is intended, the modification factor employed is 0.77.

Applying this to the permissible stress figure given in Table 3.7 (green axial stress) the allowable stress can be calculated as follows:

Allowable stress $= 0.77 \times 3.7 = 2.85$ N/mm^2

Maximum load is determined by allowable stress $\times$ area of section:

$$W = \frac{2.85 \times 425 \times 150}{10^3}$$

$$= 181.69 \text{ kN}$$

the requirement is for a total load of 350 kN, so two such sections would be needed, giving an all-up capacity of 363.38 kN.

Example 6 Timber (cantilevered) canopy (Fig. 3.7)

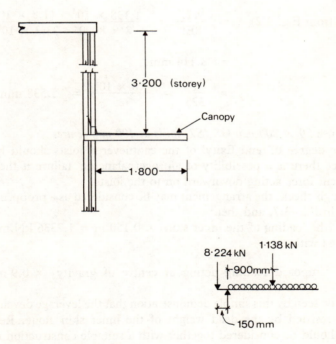

Fig. 3.7 Cantilevered canopy

A cantilevered canopy projecting 1.8 m with a built-up felt roof covering is considered in this example.

Cantilevered joists are set at 400 mm centres to suit standard board sizes.

As in earlier examples, just one joist is considered, but the 'mean' value of E and a factor to reflect short-term loading may be used.

From Fig 3.2, maximum BM is given as:

$$MA = \frac{-WL}{2} = \frac{1.138 \times 1.8}{2} = 1.024 \text{ kN.m}$$

$$Z \text{ required} = \frac{1.024 \times 10^6}{4.8 \text{ (from Table 3.7)} \times 1.25 \text{ (short term loading)}}$$

$$= 170.67 \times 10^3 \text{ mm}^3$$

taking (for example) 50 × 200 mm joists from Table 3.5:

Z actual = 333×10^3 mm³

Deflection

$$D \text{ actual (from Fig. 3.2)} = \frac{WL^3}{8EI} = \frac{1.138 \times 10^3 \times (1.8 \times 10^3)^3}{8 \times 8000 \times 33.3 \times 10^6}$$

$$= 3.114 \text{ mm}$$

$$Dp = \frac{1}{325} = \frac{1.8 \times 10^3}{325} = 5.538 \text{ mm}$$

therefore use *50 × 200 mm GS (S2) joists at 400 mm centres*
Note: The degree of 'end fixity' of the cantilevered joists should be checked, since there is a possibility of an 'overbalancing' failure if there is insufficient force acting downward on to the joists.
    As a rough check, the arrangement may be considered as a propped cantilever, as shown in Fig. 3.7, and then:
8.224 kN (the loading of the inner skin) × 0.150 m = 1.2336 kN.m
is compared with:

1.138 kN (imposed loading acting at centre of gravity) × 0.9 m = 1.0242 kN.m
and it can be seen by this simple demonstration that the leverage developed by the canopy is resisted by the dead weight of the inner skin alone. Resistance to crushing should be considered together with a suitable construction method for the support area, but this is outside the scope of the example.

Example 7 Reinforced-concrete suspended floor (Fig. 3.8)

There is almost no justification for the site engineer to become involved in on-site design calculation for reinforced concrete. Builders' merchants will usually have a local reinforced concrete manufacturer ready to supply both stock items and design schedules, usually in the form of maximum loads and spans – these

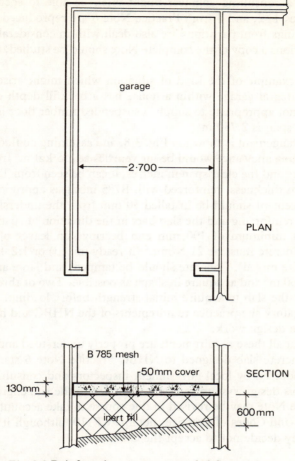

Fig. 3.8 Reinforced concrete suspended floor slab

products are normally confined to limits for UDLs of up to 30 kN (W total), but with a little ingenuity more specialised solutions are possible, and it is of course essential to obtain the supplier's written guarantee that the member is suitable for the purpose you intend. Other matters would normally be dealt with by the consultants.

One problem which sometimes arises unexpectedly in housing or light commercial construction is where the proposed fill depth below a concrete slab exceeds 600 mm, and it is necessary to reinforce the slab with steel and suspend it on the inner skin of the substructure. The NHBC (National House-building Council) has developed a complete design procedure booklet for suspended reinforced slabs, called simply: Practice Note 6 (NHBC 1973), *Suspended Floor Construction for Dwellings on Deep Fill Sites*. This is well respected and easy to use, although a consulting engineer might often be able to specify a cheaper solution. Tables 1, 2, and 4 from Practice Note 6 are reproduced as Tables 3.8 and 3.9 – loadings from partitions are also dealt with in considerable detail, but for these situations a copy of the complete Note should be studied, available from the NHBC.

A practical example of the kind of situation which might arise is where for instance, an integral garage within a house has a backfill depth exceeding 600 mm, and it is not appropriate to supply a suspended timber floor in this application, the clear span is 2.750 m.

A typical arrangement is shown in Fig. 3.8, and assuming no floor finishes, the only extra loading allowance would be for vehicles at 255 kgf/m^2 from Table 3.8. With this figure and the clear span in metres, it can be seen from Table 3.9 that a slab of 130 mm thickness, reinforced with B785 mesh, is appropriate.

The reinforcement should be installed 50 mm from the underside of the slab with the main reinforcing and the slab bars in the direction of span, and the slab should have a minimum of 100 mm end bearings on leaves of substructure brickwork. Concrete must be 21 N/mm^2 (if ready mixed) or 1:2:4 using 20 mm aggregate (if site mixed). Concrete should be tamped, and large areas avoided – no more than 60 m^2 and as square in shape as possible. Two or three days should be allowed for the slab to acquire initial strength before loading. These are the principal obligatory site practice requirements of the NHBC and have each been assumed in the design work.

Provided that all these requirements are properly understood and adhered to, a suspended concrete slab designed to NHBC Practice Note 6 standards will be confidently endorsed by local government inspectors and consulting engineers, and as such this design system can be very useful to the site engineer.

The Practice Note may be revised in due course to take account of changes in BSI standards and Codes of Practice (notably CP110) although it is understood the NHBC may decide not to accept the changes.

Tables to worked examples

Table 3.1 Properties of steel sections (usually stocked)

Nominal size	kg/m	ry(cm)	Area (cm²)	D/T	Ixx (cm⁴)	Zxx (cm⁴)
Universal beams						
356 × 171	51	3.87	64.5	30.9	14156	796.2
356 × 171	45	3.78	56.9	36.2	12091	686.9
305 × 165	40	3.85	51.4	29.9	8523	561.2
305 × 102	33	2.15	41.8	29.0	6487	415.0
254 × 146	37	3.47	47.4	23.4	5556	434.0
254 × 146	31	3.35	39.9	29.1	4439	353.1
203 × 133	30	3.18	38.0	21.5	2887	279.3
203 × 133	25	3.10	32.3	26.0	2356	231.9
Rolled steel joists						
178 × 102	21.54	2.25	27.4	19.7	1519	170.9
152 × 89	17.09	1.99	21.8	18.4	881.1	115.6
127 × 76	13.36	1.72	21.0	16.7	475.9	74.94
102 × 64	9.65	1.43	12.3	15.4	217.6	42.84
Universal columns						
203 × 203	52	5.16	66.4	16.5	5263	510.4
203 × 203	46	5.11	58.8	18.5	4564	449.2
152 × 152	30	3.82	38.2	16.8	1742	221.2
152 × 152	23	3.68	29.8	22.3	1263	165.7

Table 3.2 Allowable stress table: beams. Allowable stress (Pbc) in bending (N/mm²) for beams of grade 43 steel (from Table 3a, BS 449 Part 2: 1969)

l/ry D/T:	10	15	20	25	30	35	40	50
90	165	165	165	165	165	165	165	165
95	165	165	165	163	163	163	163	163
100	165	165	165	157	157	157	157	157
105	165	165	160	152	152	152	152	152
110	165	165	156	147	147	147	147	147
115	165	165	152	141	141	141	141	141
120	165	162	148	136	136	136	136	136
130	165	155	139	126	126	126	126	126
140	165	149	130	115	115	115	115	115
150	165	143	122	104	104	104	104	104
160	163	136	113	95	94	94	94	94
170	159	130	104	91	85	82	82	82
180	155	124	96	87	80	76	72	71
190	151	118	93	83	77	72	68	62
200	147	111	89	80	73	68	64	59
210	143	105	87	77	70	65	61	55
220	139	99	84	74	67	62	58	52
230	134	95	81	71	64	59	55	49
240	130	92	78	69	61	56	52	47
250	126	90	76	66	59	54	50	44

Table 3.3 Allowable stress table: columns. Allowable stress (Pc) in compression (N/mm²) for columns of grade 43 steel (from Table 17a, BS 449 Part 2: 1969)

l/r	P_c (N/mm²) for grade 43 steel									
	0	1	2	3	4	5	6	7	8	9
0	155	155	154	154	153	153	153	152	152	151
10	151	151	150	150	149	149	148	148	148	147
20	147	146	146	146	145	145	144	144	144	143
30	143	142	142	142	141	141	141	140	140	139
40	139	138	138	137	137	136	136	136	135	134
50	133	133	132	131	130	130	129	128	127	126
60	126	125	124	123	122	121	120	119	118	117
70	115	114	113	112	111	110	108	107	106	105
80	104	102	101	100	99	97	96	95	94	92
90	91	90	89	87	86	85	84	83	81	80
100	79	78	77	76	75	74	73	72	71	70
110	69	68	67	66	65	64	63	62	61	61
120	60	59	58	57	56	56	55	54	53	53
130	52	51	51	50	49	49	48	48	47	46
140	46	45	45	44	43	43	42	42	41	41
150	40	40	39	39	38	38	38	37	37	36

Table 3.4 Typical loadings

		Notes
Floors (timber joists)		
Superimposed load	1.50	—— Can be reduced (to
Floor structure	0.34	1.44 kN/m² for buildings
Ceiling	0.21	of less than three
	——	storeys with single
	2.05 kN/m²	occupation)
Pitched roofs		
Superimposed load	0.75	—— Plan area
Tiles/roof structure	0.90	
Ceiling	0.21	
Superimposed ceiling load	0.72	—— Access to roof void
	——	
	2.58 kN/m²	
Flat roofs		
Superimposed load	0.75	
Boards and joists	0.34	
Felt and chippings	0.28	
Ceiling	0.21	
	——	
	1.58 kN/m²	
Walls		
250 mm cavity brickwork/blockwork	5.15 kN/m²	On elevation
100 mm brickwork/blockwork	2.57 kN/m²	On elevation
Studwork partitions	0.8 kN/m²	On elevation

Table 3.5 Properties of timber sections. Timber joist properties (extract from Table 55, CP 112 Part 2: 1971)

Joist size (b × d)	Section modulus (Zxx) (× 10³ mm³)	Second moment of area (Ixx) (× 10⁶ mm⁴)
38 × 75	36.6	1.34
38 × 100	63.3	3.17
38 × 125	99.0	6.18
38 × 150	142	10.7
38 × 175	194	17.0
38 × 200	253	25.3
38 × 225	321	36.1
50 × 75	46.9	1.76
50 × 100	83.3	4.17
50 × 125	130	8.14
50 × 150	188	14.1
50 × 175	255	22.3
50 × 200	333	33.3
50 × 225	422	47.5
63 × 100	105	5.25
63 × 125	164	10.3
63 × 150	236	17.7
63 × 175	328	28.1
63 × 200	420	42.0
63 × 225	532	59.8
75 × 100	125	6.25
75 × 125	195	12.2
75 × 150	281	21.1
75 × 175	383	33.5
75 × 200	500	50
75 × 225	633	71.2

Table 3.6 Timber stress modification factors. Modification factor K_{18} for slenderness ration and duration of loading on compression members (softwood) (from Table 15, CP112 Part 2: 1971)

Slenderness ratio (l/b)	Long-term loads	Medium-term loads	Short-term loads
1.4	1.00	1.25	1.50
1.4	0.99	1.24	1.49
2.9	0.98	1.23	1.47
5.8	0.96	1.20	1.44
8.7	0.94	1.17	1.40
11.5	0.91	1.13	1.34
14.4	0.87	1.08	1.27
17.3	0.83	1.00	1.16
20.2	0.77	0.90	1.01
23.0	0.70	0.79	0.86
26.0	0.61	0.68	0.72
28.8	0.53	0.58	0.60

Table 3.6 (contd)

Slenderness ratio (l/b)	Long-term loads	Medium-term loads	Short-term loads
34.6	0.40	0.42	0.44
40.4	0.31	0.32	0.33
46.2	0.24	0.25	0.25
52.0	0.20	0.20	0.20
57.7	0.16	0.16	0.17
63.5	0.13	0.14	0.14
69.2	0.11	0.12	0.12
72.2	0.10	0.11	0.11

Table 3.7 Permissible timber stresses (extracts taken from Tables 10a and 11a, CP 112 Part 2: 1971)

Species	Grade	Not exceeding 18% moisture content (in units of N/mm^2)				Exceeding 18% moisture content (in units of N/mm^2)			
		Dry bending stress	Dry axial stress	Dry modulus of elasticity		Green bending stress	Green axial stress	Green modulus of elasticity	
				Mean	Min.			Mean	Min.
Redwood	SS	6.9	7.0	8900	4800	5.5	5.4	8000	4300
Whitewood	GS	4.8	4.8	8000	4300	3.8	3.7	7000	3900

Table 3.8 Reinforced concrete slab loadings. NHBC reinforced concrete slab design data (taken from Tables 1 and 2 to NHBC Practice Note 6: 1973) The weights are abstracted from the British Standard Schedule of Weights of Building Materials BS 648: 1964, as amended 1968 and 1969.

Weights of floor finishes (of specified thicknesses)*

Finish	Approximate weight in kgf/m^2†
Sand-cement screed, 25 mm thick	58.5
Asphalt flooring, 13 mm thick	26.8
PVC (vinyl) asbestos tiles, 4.8 mm thick	10.3
Hardwood floor blocks, 10 mm thick	7.9
Flexible PVC tiles, 3.2 mm thick	44.9

Equivalent load allowance for vehicles

Factor	Equivalent allowance in kgf/m^2† of floor area
Car parking only for passenger vehicles not exceeding 2500 kg gross weight	255

*Weights for different thicknesses, and combinations of finishes, may be calculated on a straightforward pro-rata basis. †To convert kgf/m^2 to N/m^2, multiply by 9.81.

Table 3.9 Reinforced concrete slab design. NHBC slab design showing slab thickness in mm and reinforcement mesh number (reproduced from Table 4, NHBC Practice Note 6: 1973)

Clear span of slab in (m)	Uniformly distributed load allowances for finishes and partitions, (kgf/m²)											
	97.5 (kgf/m²)		146 (kgf/m²)		195 (kgf/m²)		244 (kgf/m²)		293 (kgf/m²)		341 (kgf/m²)	
	Slab	Mesh	Slab	Mesh	Slab	Mesh	Slab	Mesh	Slab	Mesh	Slab	Mesh
2.4	130	B385	130	B385	130	B385	130	B385	130	B503	130	B503
3.0	130	B503	130	B503	130	B785	130	B785	130	B785	150	B785
3.7	150	B785	150	B785	150	B785	150	B1131	180	B785	180	B785
4.3	180	B785	180	B785	180	B1131	180	B1131	180	B1131	180	B1131
4.9	200	B1131	200	B1131	200	B1131	200	B1131	200	B1131	200	B1131
5.5	230	B1131	230	B1131	230	B1131	230	B1131	250	B1131	250	B1131

*To convert kgf/m² to N/m², multiply by 9.81.

Notes:
1. Table allows for self-weight of slab and for domestic live load required by CP3 'loading' Ch V Part 1, *Dead and Imposed loads*: 1967, as amended 1968 and 1972.
2. Clear span = clear distance between supports.
3. Concrete to have specified works cube strength at 28 days of 21 N/mm².
4. Reinforcement quoted is high-yield mesh conforming to BS 4483: 1969, as amended 1972.
5. Below the line in the table, the slab depth has been increased to permit the use of mesh reinforcement. With specially designed rod reinforcement the slab depth would be less.
6. If the actual span or load allowance in any particular situation is not shown in the table always use the span or load allowance next above. Never go below.

Table 3.10 Reinforcement data

	Main bars (mm dia)	Cross bars (mm dia)	Mesh size (mm)	Weight (kg/m²)
B385	7	7	200 × 100	4.53
B503	8	8	200 × 100	5.93
B785	10	8	200 × 100	8.14
B1131	12	8	200 × 100	10.90

Chapter 4

Setting out: general techniques

Probably the single most important function of the site engineer, and certainly that which most readily comes to mind when the term 'site engineer' is used, the setting-out process is one of:

(a) Identification
(b) Simplification
(c) Expression

where the site engineer acts as the interface between designer and workforce.

Identification

The site engineer must first understand from the drawings, and any other information provided, the designer's exact intention in all its details. It is essential at this stage to determine that every facet of the construction process is feasible, that all the levels and distances are operable, and that anything left to 'good site practice' or the 'engineer's satisfaction on site' is now determined absolutely.

Simplification

The site engineer must simplify each construction stage to a point where:

1. Verbal instructions can be remembered easily
2. A little thought will enable a foreman to deduce further information logically and accurately

Simplification is a process of translation from selected areas of the designer's plans and schedules, re-presenting the principal ideas in terms that can be acted upon on site. The site engineer must be able to express complex construction sequences in a few chalk-marks and a brief explanation.

Expression

Sufficient information must be provided in the form of profiles, nails, sand-lines, paint marks, etc, to permit accurate construction; but not so much information that there is the possibility of doubt. The moment there is confusion, the engineer has failed – there is often a temptation to give too much information too soon, and this should be recognised and avoided.

Driving in a peg

This is one of the more fundamental tasks of a site engineer and his chainman, and it is on the accuracy and efficiency with which pegs are positioned that the abilities of the engineer and his assistant are measured, and upon which the whole of the construction team relies.

Rather as an artist uses his crayon, and draughtsman his drawing-pen, the site engineer gives expression to ideas, calculations and instructions by the use of timber pegs, or more particularly, steel nails carefully and precisely mounted in the top of those pegs.

Apparently a mundane operation requiring little skill, driving a peg into the ground precisely on station and locating a steel nail on the top of it, is a complex and precise activity, and it is essential that the engineer's chainman is fully aware of the problem and its means of achievement. It would not be unusual to drive 100 pegs and nails in exact locations during the course of a working day. Before taking a new chainman on site the engineer would be well advised to arrange a practice session to enable his assistant to familiarise himself with the procedure.

Directions for driving a peg

1. Take the peg and tape 'zero' (the end of the tape) and place the point of the peg on the ground, the tape 'zero' on top of the peg, and face the engineer
2. Move the peg and tape, together, right or left as instructed by the engineer (as necessary to bring the peg on to cross-hairs on the theodolite) keeping the tape taut and the peg upright
3. When on station (site engineer will indicate by signalling with two arms extended) drop the tape and drive in the peg, making sure it is absolutely vertical, for about 50 mm
4. Pick up the tape and check the distance (the peg may have fouled a stone below ground and moved off-station) – look at the engineer for 'left', 'right' or 'on station' signals. If all is clear, proceed to drive the peg in, otherwise rectify the problem, go back to No. 2 and begin again
5. Strike the top of the peg solidly and evenly until it is firmly embedded in the ground and will not move by reasonable hand pressure, checking the tape distance once or twice on the way down. Looking full-face towards the engineer will indicate to him that an instrument check is required.

51

6. If it is necessary to adjust the peg's position slightly, there are three ways in which this can be done:

   (a) with the sledge-hammer hit the ground immediately beside the peg (this has the effect of further compaction and controlled sideways movement)
   (b) pull the peg over with one hand (kept well down) and hit it on the top with the sledge-hammer held in the other
   (c) hit the peg on the top, but with glancing blows in the desired direction – this requires much practice, but has the effect of simultaneous movement and compaction

7. Once the peg is embedded firmly and accurately on station, pick up the tape again and strike an arc with a nail held at the 'zero' (at any sensible distance the 'arc' will be indistinguishable from a straight line, but it is worth using the correct term) drop the tape and place the nail, held only by its head, on the centre of the arc.

8. The engineer will then signal 'left' or 'right' and by moving the nail very carefully in response to the signals, the exact intersection point can be found, the engineer will signal 'correct' and the nail may be driven in – usually half-way for most purposes.

9. It is then necessary to check the distance once more, and the engineer will give a final instrument check.

10. Sometimes the engineer will ask that the nail be pushed upright, into the vertical. This can normally be achieved by cautiously bracing the fingers against the peg and pushing the nail with the thumb – the peg itself must not be moved, and the nail must not be struck at this stage with a hammer, as there is every possibility of unnecessary displacement.

11. Move away from the peg when the 'all clear' sign is given by the engineer, and ensure that no instruments, tapes or other equipment can foul it before moving on.

Signalling

The engineer will frequently wish to communicate simple commands to the chainman and a few basic signs, illustrated in Fig. 4.1, can convey a great deal. The three principal signs, used and understood almost instinctively, are:

1. 'move left'
2. 'move right'
3. 'correct' (or 'yes')

Other useful signs are:

4. 'tilt left'
5. 'tilt right'
6. 'on the top'

7. 'no'
8. 'up'
9. 'down'

move left   move right   correct

tilt left   tilt right   on the top

no   up   down

Fig. 4.1 Signals

Radio telephones

It is often possible to communicate by speech (or more likely by shouting). There are times when this is not possible – it may be that a radio telephone is the answer, but for this purpose the following requirements apply:

1. The equipment should be fully portable, of 'pocket' dimensions and weight, with a short rubber-cased aerial
2. It should occupy a completely free channel for the duration of the works in hand
3. It should be capable of being permanently switched to 'transmit' for 'hands-off' communication, or have a voice-actuated transmit facility.

It is now conventional for private and commercial radio-telephone conversations to be littered with jargon, codes and abbreviations. The site engineer has no special need for this approach, nor on the other hand should long and involved discussions be conducted on the radio – the safest approach is clarity, brevity and carefully composed unambiguous messages. Think out your phrase before pressing the 'transmit' button, then speak slowly. It is often thought necessary to include 'over', 'over and out' or even '10-4' (there are many variations) at the end of a phrase, but it is in fact quite obvious to the recipient that the caller has finished transmitting, because a distinctive 'hiss–click' is heard at the moment the transmit button is released, with most equipment.

If you are unable to hear and understand the message, simply transmit 'please repeat message'. If the message still cannot be heard or understood, transmit either:

(a) 'I did not hear your full message but think you said . . . please acknowledge' or
(b) 'I did hear your message but do not fully understand – do you mean that . . . please acknowledge'

– this leaves the original caller a good chance of answering with one word only, instead of a full and involved explanation.

There may be a temptation to 'play' with the RT equipment, and whilst occasional light-heartedness is probably good for morale, continual and repetitive use of the equipment for trivial amusement must be ruled out. Profanities and obscenities must never be broadcast.

Horizontal control

Any construction site must have horizontal 'datum' or 'reference' lines which would have originally formed the basis for the design work and which can be reset on site as and when necessary for construction purposes. This is usually achieved by use of a system of 'control' points strategically positioned around the site during the original pre-design land survey, each one being related to an imaginary grid covering the whole of the site.

Larger projects will usually be 'tied in' to the Ordnance Grid (sometimes referred to as the National Grid). Site control can be 'tied in' to the National Grid using a system known as resection, relating peg position on site by geometry to known co-ordinated points (church spires, tall chimneys, lightning conductors, etc.) the grid positions for which are published nationally. It is not intended to deal with resection here as it is almost never of particular interest to the site engineer, provided that some form of reliable co-ordination is available. There are many excellent land survey textbooks which deal thoroughly with the subject.

Small projects will usually have their own 'local' grid, decided upon by the pre-design land surveyor. In such circumstances it is helpful if the 'local' grid can be oriented conventionally, i.e. with North at the top, East on the right-hand side – this can be a useful aid in running the site generally, and is particularly useful to the site engineer when working on co-ordinate calculations. The land surveyor

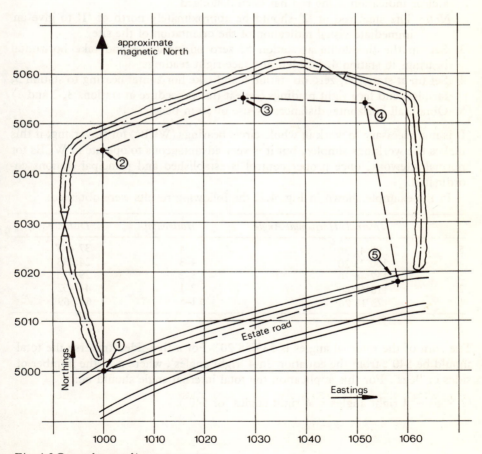

Fig. 4.2 Ground control/survey traverse

can achieve 'North' orientation by using a small magnetic compass on first arrival at the site, thereby setting the tone for the entire development.

Traverses

In Fig. 4.2 a small development site is shown and the operations required might be as follows:

1. Drive large road nails (or pipe nails) between the face of the kerb and the channel block at positions 1 and 5
2. Drive 1.2 m long (12 mm dia.) steel rods into the ground at positions 2, 3 and 4, leaving (say) 50 mm protruding above the ground. Cut a 300 mm square around each rod position, approximately 100 mm deep, and fill with concrete – this acts principally as a marker, but also helps prevent movement and gives a clear indication if the rod has been dislodged.

   Note: The steel rod at '2' should be approximately north of '1' to give an immediate visual indication of the orientation of the site
3. Set up the theodolite at station 5, zero on station 1 and take horizontal bearings to station 4, face left and face right readings
4 .Set up at station 4, zero on station 5 and take horizontal bearing to station 3, face left and face right readings (repeat this procedure at stations 3, 2 and 1)
5. Obtain all horizontal distances by use of 50 m band

It is not necessary to work in whole-circle bearings (WCBs) for traversing, if this makes the workings simpler, but it is very advantageous to operate in WCBs for normal sitework, once proper control is established and principal stations co-ordinated.

In the example shown in Fig. 4.2, the following results were obtained:

	(Internal) Horizontal Angle	Horizontal	Distance (m)
5	97°41′30″	5–4	37.475
4	100°23′20″	4–3	24.293
3	157°46′30″	3–2	29.936
2	110°48′40″	2–1	44.452
1	73°18′20″	1–5	60.569

The sum of the internal angles is 539°58′20″. For a five-sided traverse the total should be 540°, from the equation 2n-4 'right angles', where n = the number of sides or 'legs'. For this application the total internal angle should be:

$(2 \times 5) - 4$ right angles = 6 right angles (or 540°).

Adjustments

The most commonly used 'land surveyor's' method of traverse adjustment is the Bowditch Method in which misclosure distance differences are apportioned for each leg in the ratio of the length of the leg to the total length of the traverse. It is not proposed to deal with this method.

A simplified method for site engineering purposes involves equal distribution of the angular error (1'40" in the example in Fig. 4.2 with each internal angle receiving +20"). The co-ordinates are then calculated for each station using the adjusted internal angles in a clockwise direction round the traverse. A second set of co-ordinates is then calculated in the anticlockwise direction. The two sets of co-ordinates are then simply averaged to give final eastings and northings, and the distances between stations subsequently recalculated to give final distance correction. This method has no more justification than any other, but it has the advantage of being very simple to follow.

Site engineering work is generally considered to be of 'third order' in terms of land-survey accuracy. In this category permissible angular misclosure for traverses is usually taken as 150 $\sqrt{n}$ (seconds), where n = the number of sides or legs. A 20" theodolite is assumed. For the example in Fig. 4.2 the acceptable maximum using this rule is 5'35". In this case the actual closing error is 1'40". Total distance measurements are required to be within 100 mm of closure in this application.

Using the adjusted internal angles from the previous example, it is now possible to retabulate as follows:

	Adjusted internal angle	*Horizontal*	*Unadjusted distance (m)*
5	97°41'50"	5–4	37.475
4	100°23'40"	4–3	24.293
3	157°46'50"	3–2	29.936
2	110°49'00"	2–1	44.452
1	73°18'40"	1–5	60.569

Basic trigonometry and calculation of rectangular co-ordinates

The trigonometry used in site engineering is very basic but it is nevertheless worth including in this text to establish the 'ground rules'.

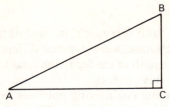

Fig. 4.3 Basic trigonometry

Sine   =   opposite ÷ hypotenuse (eg sin BAC = BC/AB)

Cosine   =   adjacent ÷ hypotenuse (eg cos BAC = AC/AB)

Tangent =   opposite ÷ adjacent   (eg tan BAC = BC/AC)

$BA^2 = AC^2 + BC^2$

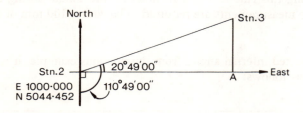

Fig. 4.4 Co-ordinate calculations

Station 3, station 2 and 'A' in Fig. 4.4 is 100°49′00″ − 90° = 20°49′00″. Distance station 2 to station 3 is 29.936 m.

The easting is calculated as:

adj   = 29.936 × cos 20°49′00″
       = 27.981 847 97

The new easting is therefore 1000 + 27.982 = 1027.982

The northing is calculated as:

opp   = 29.936 × sin 20°49′00″
       = 10.638 622 15

The new northing is therefore 5044.452 + 10.639 = 5055.091

Succeeding stations may be calculated in this way, both clockwise and anticlockwise around the traverse and the results averaged.

	Clockwise		Anticlockwise		Average	
Stn	*Easting*	*Northing*	*Easting*	*Northing*	*Easting*	*Northing*
5	1058.120	5017.479	1058.016	5017.400	1058.068	5017.439
4	1052.267	5045.494	1052.163	5054.415	1052.215	5054.455
3	1027.982	5055.091	1027.878	5055.012	1027.930	5055.052
2	1000.000	5044.452	999.896	5044.373	999.948	5044.413
1	1000.000	5000.000	1000.000	5000.000	1000.000	5000.000
5–1	1000.099	5000.096	999.896	4999.921	999.998	5000.008

The original ordinates (E 1000.000, N 5000.000) are retained throughout, irrespective of calculated closures. Using the new average co-ordinates, new distances may then be calculated. The method shown has the advantage that it is very 'visible'.

If some simple rules are followed many problems may be avoided:

1. Measure internal angles only
2. Traverse anticlockwise, backsight zero, foresight true reading (face left)
3. Read and compare 'face right' readings whilst on station
4. If angular tabulation appears incorrect do it again
5. Keep traverse legs roughly equal, maximum banding distance 50 m
6. Always close the traverse, preferably to the starting point
7. Keep traverse as 'circular' as possible
8. Take any slope correction (vertical) angles and height of instrument at the same time as traverse angles – top of chainman's head way be used as a constant and a staff is unnecessary.

Horizontal measurements

Unless the site engineer has access to EDM (electronic distance measurement) equipment, traverse leg distances will have to be obtained using a 50 m steel band – longer bands are available, but for site engineering (as distinct from land surveying) they are not always appropriate. Measurements by banding are subject to a range of potential inaccuracies, particularly over longer distances, and correction or allowances should be made for:

(a) Slope
(b) Sag
(c) Temperature
(d) Earth's curvature
(e) Scale factor

In fact (a) slope and (b) sag pose the most serious threat to accuracy in normal site engineering terms, although by far the most usual cause of problems is operator error. Simple slope corrections can be made by taking levels at each end of the band, at the measured points, and then treating the measured distance as the 'hypotenuse' of a triangle – the short side is the levels difference. By applying basic trigonometry the true horizontal distance can be found. An example is given in Fig. 4.5. Slope corrections are usually negligible for site engineering purposes if the slope angle is less than 2°.

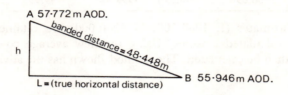

A 57·772 m AOD.
banded distance = 48·448 m

h

B 55·946 m AOD.

L = (true horizontal distance)

Fig. 4.5 Slope correction

Where $L^2 = H^2 - h^2$ in Fig. 4.5:
2347.208 − 3.334
= 48.413 m
(H = banded distance 48.448 m).

Sag correction

The need for sag correction may be virtually eliminated if:

(a) Banded distances are kept short
(b) The band is supported along its length (even flat on a road surface will do)
(c) Tension is increased (although not beyond manufacturer's limits)
but in extreme cases the formula:

$$C = \frac{W^2 L^3}{24 T^2}$$

may be applied where C is the sag correction and must be deducted from L to establish the true horizontal distance.

Example

W = weight per unit length in kg/m (0.0185 kg/m)
L = measured length (catenary) in m (31 m)
T = applied tension (via spring balance) in kg/f (7 kg)

$$C = \frac{0.0185^2 \times 31^3}{24 \times 7^2} = 0.0087 \text{ m}$$

Details to plot

The site engineer may sometimes be asked to carry out a pre-contract land survey to obtain and check important contract information.

The engineer is fortunate in that unlike more general land surveyors he knows in advance what information is relevant and what is not. This makes his job rather easier than might be the case. He might be interested in, for instance:

- Drains
- Water, electricity, gas and telephone
- Trees
- Levels
- Overhead cables
- Ground surface conditions
- Perimeter roads and accesses
- Topographical details
- Fences
- Gates
- Footpaths and rights of way

but not very concerned with

- Existing structures on site
- Topographical details to remain unaltered

– because this is general survey data that are available already from a variety of sources, and would already have been taken into account.

On a more general basis, the site engineer will be especially concerned to discover the actual site dimensions and bring to light any disputed boundaries, indistinct ownerships and other problems which can affect the dimensional co-ordination of his project.

Drains

The exact positions of inspection chambers, manhole covers, gullies, etc. should be determined and referenced, and both cover and invert levels must be recorded.

Special equipment may be needed to open the chambers, but the job must be done.

Water, electricity, gas and telephone

Service undertakings will usually supply a detailed plan showing their plant and installations, including an indication of depth – although they will not guarantee accuracy. If the exact position of these items is critical to the proposal, trial excavations may be necessary to establish the precise situation. This is best done by hand, with a representative of the appropriate authority in attendance. Sometimes service runs can be located by electric metal-detectors. This can work well for electrical and telephone equipment, but not always for water or gas supplies if they are in plastic pipes; Some people can detect water and various minerals below ground by divining – this is not a well-understood accomplishment, but is worth a try if there is no definite information available.

Trees

As already mentioned in Chapter 2, to the site engineer all trees are a physical obstruction, and a potential nuisance. To allow the various official bodies to consider an application for felling and removal, a 'tree survey' will often be required, and this usually takes the form of:

(a) An accurate plot of each tree position on a 1 : 500 scale plan of site
(b) An identification label indicating the species
(c) Details of
    (i) girth (the circumference of the tree at a point 1.3 m above the ground) and
    (ii) spread (the approximate diameter of the tree from its trunk centre)

Occasionally the site engineer may be asked to determine the height of a tree – because of its general unevenness and large bulk, it can be quite difficult to do this precisely, especially for a deciduous tree in windy conditions, but a method is shown in Fig. 4.6.

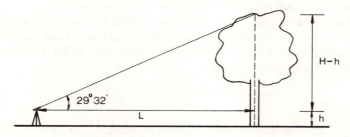

Fig. 4.6 Tree height

The distances L and h are obtained by taking readings on a staff set at the base of the tree (L by tacheometry, h by levelling the theodolite's vertical circle and taking a normal 'level' reading). H−h is obtained by:

$$\tan 29°32' = \frac{H-h}{L}$$

if h = 1.42 and L = 42.25,

$$H-h = 42.25 \times \tan 29°32'$$
$$= 21.139 \text{ m}$$

therefore: H $= 21.139 + 1.42 = 22.559$ m

The engineer would be wise to round off this figure to the nearest 500 mm, and state 'approximately' when reporting the result, in the case of a tree.

In a situation where only a few trees are to be surveyed, simple taped measurements taken along and from boundaries will be adequate. Where a great many trees are involved, the best approach is probably to carry out a detailed tacheometric/angular measurements survey based on a simple set of traverse points. It is essential that every tree be allocated a reference number. This should be clearly marked on a piece of card and fixed to the tree where it can be seen from the intended direction of the instrument positions.

There are three basic groups of trees that are of interest to the site engineer:
'coniferous' (cones and needles, softwoods)
'deciduous' (broad leaves, typically hardwoods)
'fruit'   (food-producing)
– there are, of course, exceptions in each of these cases. There are many excellent books dealing specifically with tree identification, and a pocket-sized edition can be very useful in such an exercise. Trees may be identified by:

(a) Leaf shape
(b) Twig and bud configuration
(c) Seedling shape
(d) Bark texture and pattern
(e) Overall shape and form

– and recognition can become immediate with practice.

Levels

The basis of all levelling is that the height of any **given** point can be compared with the height of a known datum point. If there is no convenient datum, the site engineer may select some distinctive permanent feature and give it a nominal value – the best choice for this is usually a manhole cover to a main sewer, preferably sited in an existing road construction that will not be altered. It is also likely with this kind of feature that the true datum level is recorded somewhere and can later be checked.

Figure 4.7 shows a small levelling exercise, and Figs. 4.8 and 4.9 illustrate how it may be 'booked' using the two alternative techniques:

(a) Collimation
(b) Rise and fall

– the collimation technique is now more widely used, mainly because there is slightly less arithmetic to handle and also that modern electronic equipment (survey and calculation) finds it convenient. There is, however, the problem that none of the intermediate sights can be checked arithmetically within the booking system, as only change points are considered, whereas the rise and fall method takes every entry into account.

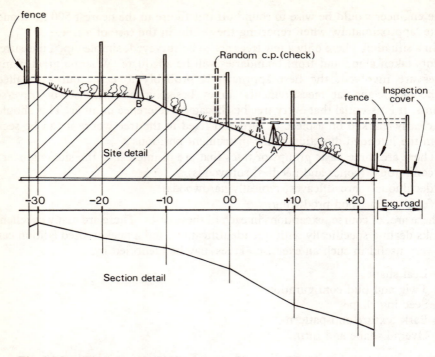

Fig. 4.7 Levelling exercise

BS	IS	FS	Coll	RL	Remarks
3·824			53·824	50·000	*INST. @ 'A' (MH, EXG RD)*
	3·620			50·204	*+ 22·8 (boundary)*
	3·304			50·520	*+ 20*
	1·795			52·029	*+10*
4·280		0·240	57·864	53·584	*INST. @ 'B' (C.P.)*
	2·396			55·468	*−10*
	1·634			56·230	*−20*
	0·590			57·274	*−30*
	0·610			57·254	*fence − 31·4*
0·465		4·210	54·119	53·654	*random CP*
		4·112		50·007	*INST.@ 'C' (MH, EXG RD)*
NOTE:	*7mm closing error*				

Fig. 4.8 Level booking: collimation

BS	IS	FS	Rise	Fall	RL	Remarks
3·824					50·000	INST @ 'A' (MAExe E)
	3·620		0·204		54·204	+22·8 (boundary)
	3·304		0·316		50·520	+20
	1·795		1·509		52·029	+10
		0·240	1·555		53·584	INST @ 'B' (CP)
4·280						—
	2·396		1·884		55·468	−10
	1·634		0·762		56·230	−20
	0·590		1·044		57·274	−30
	0·610			0·020	57·254	fence −31·4
		4·210		3·600	53·654	random CP.
0·465						—
		4·112		3·647	50·007	INST @ 'C' (MAExe E)
TOTALS						
8·569		8·562	7·274	7·267		
−8·562			−7·267			
·007			·007			
7 mm	closing error					

Fig. 4.9 Level booking: rise and fall

It will be noted from the tabulations in Figs. 4.8 and 4.9 that a levelling error of 7 mm is discovered, and it must be decided if this is acceptable in relation to the project in hand. It is not possible to be too categoric here, but tolerances are usually a matter of common sense. In the example given, taken from a cross-section detail for a new road excavation at the primary earthworks stage, 7 mm of levelling error is of no consequence. At the final 'blacktop' stage, however, levelling errors of 7 mm ± can make a great deal of difference to the final cost and may even result in a noticeable unevenness in the finished road surface.

Overhead cables

Overhead electrical 'grid' distribution cables carrying very high voltages are often the cause of considerable concern on site, and there are many instances of electrocution due to construction machinery being brought too close to power lines. Actual contact is not necessary – electricity will 'jump' or 'arc' over a distance of several metres in favourable conditions.

Cables suspended between pylons are tensioned so that sufficient free movement is possible to accommodate changes in length due to changes in temperature. The cables take up a natural catenary, and therefore little reliance can be placed on measurements of the cable height at pylons.

The site engineer will usually want to know the minimum clear height of cables immediately above any given construction operation, and this is easy to calculate by means of the technique described for measuring tree height, and as shown in Fig. 4.6. Care must be taken to identify the *lowest* cable with the theodolite telescope, and to make allowance for wind movement, which can be considerable.

Ground surface conditions

The factor most commonly determining ground surface conditions is water in the soil. The site engineer should be able to make specific recommendations concerning land drainage for the purposes of construction work, and indeed should be able to state if the ground conditions are such that stormwater soakaways are likely to be sufficient for the finished construction, or if a surface-water sewer system is necessary.

The first consideration is to establish the natural level of the water, if any, below ground – termed the 'water table'. If water is found naturally very close to the surface, there is every likelihood that the area will flood in a rainstorm, and the ground conditions might be generally boggy or swamp as a norm.

Cohesive subsoils of the clay and silt variety are generally impermeable, seriously restricting water flow. Non-cohesive soils of the gravel and sand type are much better at allowing water to escape, and therefore when found above natural water-table level are often dry within minutes of a rainstorm.

The depths, condition and outfall of any ditch on site may prove critical to the progress of the contract work – it may be that a ditch has not been cleared for thirty years or more, and that for the sake of a few hours' work with a machine, many weeks of delays can be avoided.

Perimeter roads and accesses

It is important that perimeter roads be kept clean, as mud brought on to existing roads can cause a hazard to passing traffic.

Cleaning an existing highway is a time-consuming and expensive operation, which can be rendered unnecessary by proper arrangements and planning on site. Items which should be considered are:

(a) Temporary gravel roads on site between public highway and any storage/ compound area
(b) Gravel-surfaced storage area
(c) Wheel-cleaning brushes and high-pressure hose available at site exit point.

The width, construction and surface drainage gradients of any existing perimeter road should be investigated by the site engineer at the outset, and he should be satisfied that it is:

(a) Of adequate width to allow lorries to turn into highway (otherwise a 'slip' road must be constructed on site)
(b) Of adequate strength of construction to resist the additional rigours of site traffic for the duration of the contract
(c) Drained such that rainwater does not flow into the site or collect at the access point.

At a new access point, any new or existing services or drains should be checked for line and level, and any protection or lowering arranged as appropriate. Telephone cables can be 'slewed' (moved laterally) to a limited extent; water, gas and electricity services can be lowered by the appropriate authorities (mains water will flow uphill as it is under pressure), but generally drains cannot arbitrarily be laid at lower levels without redesigning the whole system.

If a ditch crosses an access point, a culvert should be provided, even if there does not appear to be any water flowing at the time. Normally a few lengths of 225 mm or 300 mm diameter concrete storm sewer pipe will be adequate, surrounded in concrete

Site access roads should ideally be the eventual new 'estate' roads for the development, as this saves unnecessary road-building. There will be stages, however, when access is not possible along the new roads layout, and a temporary road is required. This should be built in *exactly* the same way as any other road, the topsoil stripped, a good formation exposed and a good-quality sub-base material laid. There is little need for blacktop on a short-duration site, but for a long-term development a minimum blacktop base course should be laid, with proper attention to storm drainage.

The reasoning is that if a roadway is required, then it must perform to normal roadway specifications for the duration of the requirement, and a 'cheap' access road almost always becomes expensive in the long run.

Topographical details

The site engineer has, at the 'production' stage, almost no use for contour lines, but often finds that the consultants have used low-quality information of this kind exclusively in the design stage, probably when no other data were available. Final topographic information is then obtained by the site engineer by means of cross-sections through relevant parts of the site, when the depths and heights of all inter-active elements can be shown on the same drawing. The site engineer often has considerable responsibility for interpretation of the designer's requirements in the light of site conditions.

Setting up a theodolite
Some site engineers find much difficulty in acquiring speed and precision in setting up an instrument, but if a proper sequence is followed, the problem is much simplified (See Fig. 4.10):

- Hook the plumb-bob under the instrument and centralise the instrument and legs over the station, with the baseplate approximately level
- Press the foot of each leg firmly into the ground, and check all screws for tightness (except for the fine-adjustment screws on the instrument head)
- Adjust the leg lengths until the plumb-bob hangs exactly over the station point

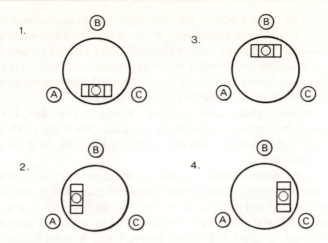

Fig. 4.10 Levelling a theodolite

- In position 1, rotate both screws A and C in opposite directions to centralise the bubble
- In position 2, rotate B to centralise the bubble
- Return to position 1, rotate both screws A and C to centralise the bubble
- In position 2, rotate B to centralise the bubble
- Repeat levelling at positions 1 and 2 until no further adjustment is required
- In position 3 note the *exact* distance by which the bubble is off-centre, and rotate both A and C in opposite directions to *halve* the distance
- Return to position 1 to check that the *halved* off-centre distance is the same in this position, if not, go back to position 3 and try again, then compare once more with position 1
- In position 4 rotate B to give the bubble the same off-centre distance as in the other positions, and check with position 3
- Repeat until there is no noticeable movement at any point.

At this stage the instrument can be said to be correctly levelled, but it may be off-station. Most modern theodolites have an 'optical plummet' and this should now be used to determine that the instrument is exactly on station.

If it is not, the baseplate may be loosened, the instrument moved directly over the station and the baseplate clamped tight once more.

The instrument may now be relevelled, but should require only a very small amount of adjustment.

An experienced site engineer will know the mean position of the bubble in his theodolite, or will have adjusted it to the centre markings, and in practice will spend only sixty seconds or so levelling the instrument. A trainee may take over thirty minutes initially, but with practice this should soon come down to a more workable five to seven minutes.

This procedure is all very well in good weather conditions, but in even the lightest wind it can prove impossible to use a conventional plumb-bob.

Some theodolites are equipped with a telescopic centre-rod incorporating its own bubble, but many have only an optical plummet, which can be used indirectly by trial and error to locate the instrument in successive approximations over the correct point – the whole instrument requires moving and resetting between each use of the plummet.

There is also a 'direct' method of levelling a theodolite optically, and although this requires (if anything) more skill and practice than the 'plumb-bob' method, this is justified by the fact that the instrument can be set up in very adverse conditions:

- Centralise the instrument and legs over the station, with the baseplate approximately level, and press the foot of each leg firmly into the ground – check all screws
- Viewing through the optical plummet, rotate the levelling screws until the station is exactly in the cross-hairs
- Using the theodolite *leg* adjustments only, level the bubble on the baseplate (the same sequence as the first example, using the positions shown in Fig. 4.10)

It may take the beginner a little time to acquire 'feel' for the delicate movements necessary, but many first-class site engineers use only this technique, irrespective of weather conditions.

Tacheometry

Distances can be measured optically using a theodolite's stadia hairs and a staff – the difference between the upper and lower readings multiplied by (usually) 100 gives the horizontal distance. The measurements obtained may be taken as accurate to ±100 mm, and this is quite acceptable for, for instance, infill survey detail, level grids, etc. where the product is a small-scale drawing and such inaccuracies are lost or tend to be cancelled out.

Figure 4.11 shows a standard arrangement for 'cross' and 'stadia' hairs in a theodolite.

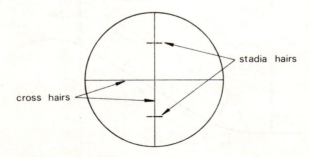

Fig. 4.11 Stadia hairs

For the example in Fig. 4.12, readings on the staff of 1.982 m for the top stadia, and 1.658 for the bottom, are taken.

The horizontal distance (H) would be:

H = (1.982 − 1.658) × 100
   = 32.4 m

– the middle (cross-hair) reading should be 1.820 and it is useful to book 'mid' readings as a check.

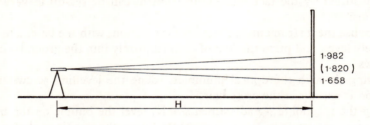

Fig. 4.12 Tacheometric distance measurement

Inclined sights

Horizontal measurement Stadia tacheometry is useful in quickly obtaining true (within ±100 mm) horizontal distances from inclined sights. In the example shown in Fig. 4.13 the formula:

H = $x(100 \cos^2\theta)$

is applied, where

H = horizontal distance
x = top stadia reading – both stadia reading
θ = angle of inclination

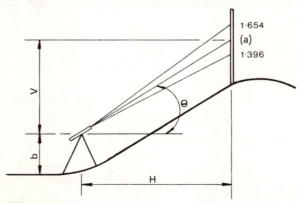

Fig. 4.13 Inclined tacheometry

70

Given that the two staff readings are 1.654 and 1.396 then

x   =   1.654 − 1.396
     =   0.258

– and given the θ = 12°15′ (cos 12°15′ = 0.977 231)

H   =   0.258 (100 × 0.977 231^2)
     =   24.639 m

Vertical measurement The vertical height V may also be found by similar means, using the formula:

V   =   x(100 cos − sin θ)

where V   =   vertical distance (less a, plus b)
      x   =   top of stadia reading – both stadia reading
      θ =   angle of inclination

For the same values as before:

V   =   0.258 (cos 12°15′ × sin 12°15′)
     =   0.258 (100 × 0.977 231 1 × 0.212 177 7)
     =   5.349 542 7
     =   5.350 m

If height of instrument (b) = 1.455 and mid-staff reading (a) = 1.525, actual level difference is:

V − a + b = 5.350 − 1.525 + 1.455 = 5.280 m

Chapter 5

Setting out: buildings

The major site engineering effort in most buildings occurs at the 'foundations and substructures' stage, and usually by the time a building has begun to emerge into daylight the site engineer's principal tasks have been over for some time.

From the engineer's point of view there is very little difference between a semi-detached pair of houses and an oil refinery complex – the techniques employed are just the same. Buildings are set out in two stages:

1. Determination of principal points for construction
2. Preservation of principal points by use of profiles

House foundations

In Fig. 5.1 a pair of semi-detached houses on an 'infill' plot with existing

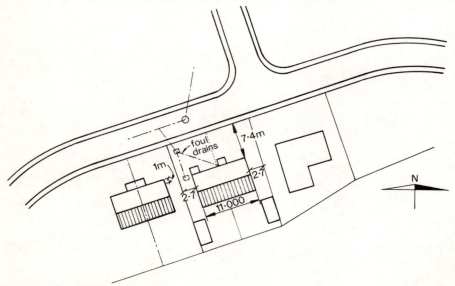

Fig. 5.1 Typical site plan

boundaries is shown on a site plan (1 : 500 scale). The forward corners of the building should be set up first, the correct distance apart, 7.4 m from the back of footpath as shown on the drawing and set 1 m back from the adjacent property.

The distance across the site should be measured and it should be checked that the 2.7 m (min.) figure is obtainable on both sides of the structure – if there is more, it is probably better to add half to each side instead of only one side, but this should be mentioned to the designer before proceeding.

The distance down the depth of the site should be checked against the plan, as it is better to find inaccuracies and discrepancies now rather than later – a near boundary encroachment might be suspected and should be dealt with as quickly as possible.

Drains and services positions should be shown on the site plan – copies should be given to the various authorities concerned and the information transferred to a single 'services' print. Details of new and existing services installations will be important to the groundworks supervisor, to avoid accidental damage and allow for excavation of new service trenches in the correct places.

Having determined the principal points from the approved plan, it would be appropriate to have the local authority planning officer check the siting – this may be delegated to the building inspector. If there is the slightest problem concerning proper siting, confirmation by letter to the council offices should follow any agreement reached on site.

Defining the outline

When the forward line is agreed, it is relatively straightforward to set out the remaining principal points using one of several pieces of equipment:

(a) Timber 'square' constructed on site as a giant right-angled triangle – cumbersome but reasonably effective
(b) An optical 'site square' commercially produced as a site tool, with a fixed 90° angle between two small telescopes, each with a 'V' sight – quite accurate but limited to a couple of basic functions and of low optical power.
(c) A surveyor's level fitted with a 360° horizontal circle and possibly a vernier attachment – accurate, with a powerful telescope, but limited by having its telescope fixed on the horizontal plan which could prove unsatisfactory on a slightly hilly site
(d) The Pythagorean '3–4–5' triangle (or its derivative '7.650–11.000–13.399' in our case for Fig. 5.2) which can be very effective, particularly if setting out alone, provided two tapes are available
(e) A theodolite

Tolerance

Figure 5.2 shows four pegs, set at the main external brickwork returns. Any of the instruments mentioned in a–e above will be capable of giving a result to

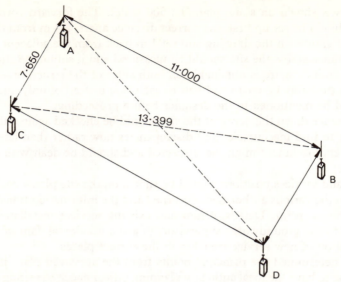

Fig. 5.2 Defining the outline

within ± 3 mm and this tolerance will be very acceptable to all parties for works of this type, as the various trades involved at this stage would not normally be capable of tolerances better than ± 12 mm.

It is essential for the site engineer to provide information of a higher order of accuracy than can be achieved by the tradesmen. The tolerance limits can be additive – the +3 mm from the engineer added to the +12 mm from the bricklayer will result in a building 15 mm too long, an increase of 0.136 per cent. This may not matter in a conventional brickwork structure, indeed it is almost to be expected, but in a situation where (for instance) system-construction panels are to be bolted together on the substructure, a cumulative error of this type may prove more critical, and in such a situation it may be to everyone's advantage to work to tolerances set by the manufacturer.

'Out of square'

It is possible to inadvertently set up an out-of-square 'rectangle' where the external dimensions are correct, but the diagonals are different from each other. In this situation the engineer should first recheck *every* external dimension, and if no serious error is found, proceed with correction of the parallelogram to a true rectangle. In Fig. 5.3 the forward line AB is the 'control' – set up as the front of the building and agreed for position. Subsequent pegs are therefore the only ones to be moved. It is possible to calculate precisely the amount of shift required to correct inaccuracy, but it is usually sufficient to move C and D by *half of the difference between the two diagonal measurements*, on a line parallel with AB and *in the direction which increases the length of the shorter diagonal.*

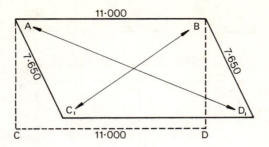

Fig. 5.3 'Out-of-square' correction

Generally the amount of movement required will be very small, and it may be possible to adjust only the nail position – in this case the 'old' nail should be hammered down flat into the top of the peg and a new nail set up beside it. Clearly, both pegs C & D must be moved by the same amount. If the peg itself must move, the possibility of tilting it should be considered, ensuring that the sole direction of movement is as required. It is possible for all manner of errors and problems to be introduced in this process, and it is better to avoid the necessity for 'out-of-square' corrections by employing techniques which make such a situation unlikely, or indeed, impossible.

Secondary details

The site engineer will not normally deal with pegs and profiles for individual details of porches, internal walls, chimneys, protrusions, etc. as these can be dealt with easily and efficiently by direct measurement by the various tradesmen who will take the engineer's principal lines as their main reference. These 'secondary details' should not simply be dismissed as unworthy of attention – the reason for omitting them is that too much detail can be very confusing, and tradesmen will usually prefer to measure, square off and set out their own secondary details within a framework provided by the engineer.

Preservation of principal points

Because the pegs and nails already set out at the principal corners of the building are, by definition, in the way of the construction work to follow, arrangements must be made so that the corner points can be re-established quickly and accurately, and this is usually done with off-set profiles. Figure 5.4 shows a total of 20 pegs, 10 boards and 50 nails (assuming 2 per peg and 1 per board) to preserve the 6 principal points of the building setting-out.

It should be noted that midpoints on the principal lines are centre-line points, representing the centre of a cavity (or solid wall) and not face brickwork.

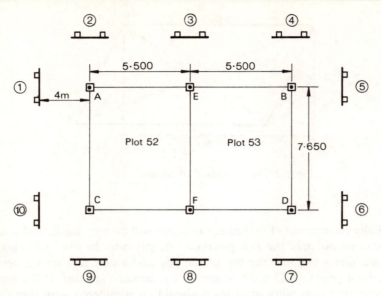

Fig. 5.4 Profiles for semi-detached houses

Sequence of operation

1. Place four pegs and two boards by each corner peg, two pegs and one board by each centre-line peg, fifty nails in pocket
2. Stand both pegs together and by sighting, place them in line with, for instance, AB at a minimum distance of 4 m from A
3. Separate the pegs, equidistant to each side of the sight-line, and hammer them into the ground until they are firm. They do not both have to be at the same level.
4. Nail the profile board to the front of the pegs, tap 'profile' nail lightly into the end-grain of one of the pegs for use later and move on
5. Repeat this process a further nine times
6. Take a 'bricklayer's' line, tie a loop in one end and place it over, for instance, the nail at peg B
7. Pull it taut over the nail at peg A and over the profile board at profile 1
8. Determine that the line is directly over the nail. Any differences in level, causing the line to foul, may be overcome using long spirit-levels.
9. Mark the profile nail position with a pencil and tap the nail home so that around 20 mm of shank remains visible
10. Repeat this process a further nine times

Preparing for excavation: sand-lines

Groundworks crews have almost no use for profiles (they are needed later on), but will use the principal setting-out points as a basis for 'sanding' their excavation

Fig. 5.5 Sanding excavation centre-lines

centre-lines. Sand is used because its colour is generally in contrast to the soil, and a clean even line can be achieved quickly – the survival period is short so sanding should be carried out immediately prior to excavation.

Sanding should be done under the direct supervision of the engineer, who can continuously set up lines and issue instructions, but fully involving the ground-works crew in actually sanding the lines.

This gives the opportunity for instant problem-solving as the work proceeds, makes the ganger familiar with the layout in the space of a few minutes and allows the excavator operator (or 'digger driver') to work out his plan of attack so that he is not left with an 'island' in the middle of an excavated area.

Centre-lines

The engineer must ensure that only dig *centre*-lines are sanded – the principal points set out earlier are for external face brickwork, and must be inset by, for instance, 125 mm (for a 250 mm construction) to reach the centre-line. Chaining arrows can be used to define the points in the ground, and the stringline can be tied off in temporary loops to allow it to be stretched taut.

Sand should be taken from a bucket and placed carefully along the stringlines by the groundworks crew (see Fig. 5.5). It will not be possible to string out every line in advance, so there will inevitably be much activity for the duration of the sanding process – it is up to the site engineer to think ahead continuously, throughout the procedure.

All internal partitions (where these are carried on foundations) chimney footings, porches, etc. must be accounted for. Generally, the line and width of the

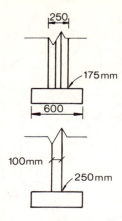

Fig. 5.6 'Spread' on foundations

main external walls will be the most critical, as these are in the majority and have comparatively little tolerance – Fig. 5.6 shows a maximum of 175 mm 'spread' on a conventional 600 × 225 mm strip footing. For reasons of expediency the groundworks crew is unlikely to change to a 500 mm trench-width for just a few internal walls, and therefore the available 'spread' here is up to 250 mm.

Ideally, the minimum 'spread' on either side of a cavity wall should be 150 mm.

Levels

Concrete levels will be determined by, particularly, the bearing surface level, correctly termed the 'formation' level (referred to as the 'bottom'). Having satisfied the building inspector and consultants that a 'good bottom' has been reached (i.e. that a suitable bearing stratum has been exposed), the excavator operator will be instructed to maintain the same level for the remainder of the exercise.

It often happens that the subsoil strata slope naturally, and it is usual in these cases for the building (or buildings, e.g. a terrace of houses) to be 'stepped' at intervals so that it follows the natural contours of the terrain. Whilst this will have been decided upon at a very early design stage, it is up to the site engineer to ensure that foundations are laid to suit both the designer's requirements and the subsoil conditions.

In the case of a major structure of immense weight, a consulting engineer will probably have carried out test borings and had soil samples analysed – he will then have been in a position to design precisely dimensioned foundations, supported by calculations. In the case of two-storey domestic housing, however, the design of foundations and their bearings is largely an intuitive process conducted on site by the site engineer and the building inspector – aided of course by the various statutory regulations, recommendations and codes of practice, which will have been incorporated into the contract drawings and/or specification, and which will) define minimum widths, thicknesses and mix specifications.

On completion of the main bulk of the excavation, which will be controlled largely by the groundworks ganger, the engineer will be expected to provide final levels for the prepared formation level, using a dumpy and a levelling staff (carried by chainman or a groundworker). In the case of a flat site the procedure is as follows:

1. Take levels on the bottom of the trench at the corners (A, B, C, D) and find the lowest point
2. Inspect the remainder of the excavation to check that no seriously lower spots exist
3. Have chainman/assistant drive first steel level pin (6 mm dia. MS rod 600 mm long) into ground at the lowest point. Theoretically the pin should be hammered into the ground until exactly 225 mm is left protruding, and then this 'control' top level maintained on all the level pins around the excavation. In practice, to balance concrete costs with labour savings, the engineer may decide to take the average of highest and lowest levels as the 'control' level, which will result in more concrete being required, but less labour-intensive shaping of the 'bottom', with the concrete varying in thickness from 225 mm (as the minimum) to as much as 280 mm. The decision must take into account the site conditions, and becomes easier with experience
4. Using the top of the 'control' pin as a datum level, place further pins at 2 m centres around the excavation and level each one – adjust by striking with a *hammer* (not using the bottom of the levelling staff) or by extracting if too low
5. At each level pin the depth should be checked and the surplus formation material removed with a clean shovel. This process is known as 'bottoming up' and its purpose is to ensure a flat bed and adequate concrete thickness.

If the site is sloping, stepped foundations will be necessary. Figure 5.7 illustrates a 225 mm step, and it should be noted that the concrete is double thickness over a length of 600 mm. The step (often referred to as a 'jump' on site) is formed using a 225 mm × 25 mm board, probably of plywood. This is usually driven into position with a sledge-hammer, supported and braced in position with pegs driven into the side-walls of the excavation.

Procedures in construction

Concrete pour

The concrete pour must be slow and even, with the mix dropped only a short distance – a vigorous drop will separate aggregate from cement and this will adversely affect the performance of the concrete. It is important that only the correct quantity of water be added to the mix – extra water will make the concrete

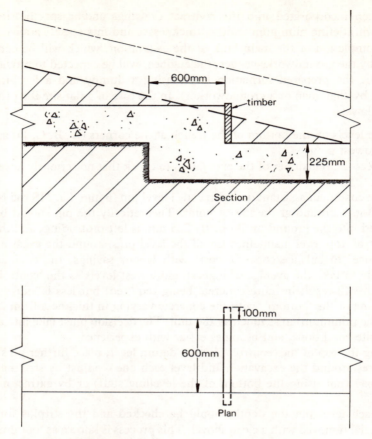

Fig. 5.7 Steps in foundations

run more easily around the excavation, but it will also reduce concrete strength (water/cement ratio) and can encourage separation of the aggregate from the cement.

Substructure brickwork

At least seventy-two hours should be allowed for the foundation concrete to harden, then substructure brickwork can commence. The original location pegs will have disappeared during trench excavation, but the profiles should still be in good condition.

Given a reasonable set of profiles, most bricklayers will be capable of setting out internal walls, porches, etc. by measurements taken from the drawings, to within ± 5 mm accuracy.

Instructions and assistance will be needed on the intended DPC/floor levels, and the site engineer can supply this information using normal levelling techniques, by reference to the TBM (Temporary Bench Mark) and the agreed floor level. It is likely that the bricklayer will wish to alter the DPC level slightly, to accord with the nearest full course of bricks (often referred to as 'to work courses' on site) and provided this is not too dramatic, a minor alteration is permissible – remember that it is usually better to be slightly higher than agreed, than slightly lower.

Indicating DPC levels

The clearest and simplest method is to provide a peg at each major return in the brickwork, and mark each with a pencil. The tops of all the pegs can be levelled and the distance down from the top to the finished level can be calculated, measured and marked, or pencil and levelling staff can be used together for more rapid results. Bricklayers build corners first as a rule, and it is the position and level of the corners that dictates the line and level of the infill work to follow.

For some types of construction it is vital to obtain an exactly uniform DPC level, particularly for timber-framed construction, but a tolerance of ±10 mm is normally acceptable. The site engineer, of course, works to tolerances of ±3 mm.

Superstructure

A storey-height rod (or 'gauge rod') should be used on site to determine the number and spacing of brick courses (the bed–joint thicknesses can be varied to adjust overall height of a given number of courses) – the site engineer should not provide this, but should ensure that the foreman bricklayer has the correct dimension between DPC level and underside of floor joists for the floor above.

If incorrect, the staircase may not fit, and/or the room height may be below the minimum regulation height.

The superstructure will usually require virtually no major setting-out because its position and dimensions are fixed by the substructure. It is, however, a good idea to check the size and shape of the finished substructure whilst still at DPC level, as any defects can be dealt with more easily at this stage than later on.

Steel-framed buildings

Although in principle exactly the same as setting-out a pair of semi-detached houses, a steel-framed building usually has a great number of 'principal points' – the centres of column bases at frequent intervals – and it is more appropriate in most situations to set out using only a theodolite and steel tape, due to the greater size of the enterprise and the necessity for accuracy over relatively long distances.

In a steel-framed building, the principal columns should be installed perfectly upright and the intermediate columns may then be set off from them. Probably the least cumbersome technique is to use two theodolites, set at 90° to each other as shown in Fig. 5.8.

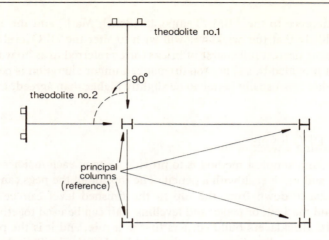

Fig. 5.8 Upright installation of steel columns

The profiles will originally have been set up for centre-lines of columns, and for this exercise new nails should be set in the profiles, representing the extremities of the column. Theodolites should be placed over the profile boards and levelled, zeroed and checked before the column descends into view. If the site engineer has only a chainman or tradesman to operate the other theodolite, it may take some time to synchronise the instructions (in this instance the engineer will of course set up both theodolites, and also check them afterwards) – the site agent/project manager/quantity surveyor may be reasonably conversant with the instrument and might be available briefly to assist.

Instructions are needed by the crane operator initially, and then by the steel fixers. Provided that the column locates readily on its HD (holding down) bolts, the procedure is one of signals to the steel fixers and adjustments to temporary supports fixed between the column and the next baseplate positioned. When correctly aligned from both viewpoints and rigidly fixed by temporary ties, the baseplate voids can be filled with a rich grout and left to harden.

Piling

Again, the basic techniques are just the same – there are various types of piles, but the site engineer will normally have little to do except set out the works prior to commencement, and then stand back. Most piling is carried out as a 'supply and fix' specialist subcontract requiring only limited help from the main contractor. Most pile arrangements for building foundation purposes support a 'ground beam', which is usually a steel-reinforced concrete beam case *in situ* over the pile caps and the superstructure can be built off the beam just as if it was a strip foundation. There are variants, but the engineer's interest will initially be confined to the centre-lines.

Pile tests

The site engineer may be involved on a purely *ad hoc* basis, in pile tests when specified by the consulting engineer from time to time. Typically, the pile selected for performance test is loaded with many tonnes of concrete weights and a series of movement-gauge readings are taken at intervals varying between ten minutes and one hour, for (typically) twenty-four hours. Then the load is increased and a further set of readings obtained. Finally all the load is taken off, and a third set of readings taken. Sometimes the consultants may specify a longer-term test, lasting a week or more.

It should be noted that the site engineer often has no direct responsibility or authority in connection with the work carried out by the specialist piling contractors except, of course, to liaise with the specialist's site engineer on basic setting-out, site services, etc. The main contractor's interests have to be looked after, however, and the site engineer's task is to keep a general watch on the proceedings – probably the best way to handle the situation is to take a genuine and lively interest.

Drains

As in other areas, the site engineer is principally concerned with communicating essential facts about a proposed drain run to the drainlayer – he is not concerned with design of drain systems, or specification of materials. There are, however, some basic requirements to consider:

1. All drainage measurements (vertical) are based on the invert level, the lowest point on the inside of a pipe – see Fig. 5.9.
2. Manholes/inspection chambers/rodding accesses, etc. are usually required at every change in direction and/or level, or at intervals not greater than the longest available cleansing equipment.

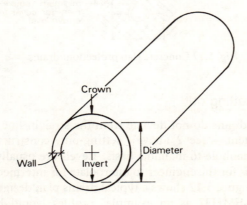

Fig. 5.9 Drain pipe terms

3. A fall is required 'across' an inspection chamber, i.e. the channel must be set at the same gradient as the drain runs on either side
4. All drains should be capable of withstanding tests (air, water, etc.) to CP301 unless they are specifically designed as perforated drains according to the contract documents
5. No drain should run within the bearing area of a foundation, taken as a 45° line from base of footing (see Fig. 5.10)

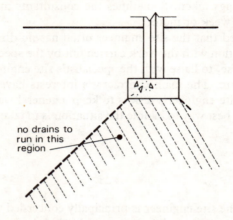

no drains to
run in this
region

Fig. 5.10 Bearing area of foundations: drains

6. Drains running close to surface and likely to be damaged or dislodged to be protected by concrete slab over-spanning the trench (see Fig. 5.11)

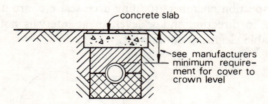

concrete slab

see manufacturers
minimum require-
ment for cover to
crown level

Fig. 5.11 Concrete slab protection: drains

Procedures in setting out

On the basis that drains do not deviate from the straight line (in terms of either direction or gradient – see 2 above), setting-out, construction and testing is carried out on a manhole-to-manhole basis. Levels are usually specified at manholes only, and it is for the engineer to interpolate for intermediate levels, usually at 5 m intervals. Figure 5.12 shows a typical drains plan design detail, and taking the run FMH36–FMH37 as an example, profiles would be erected at 5 m intervals as shown in Table 5.1.

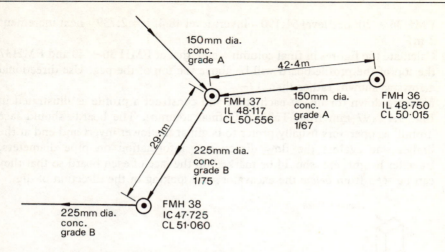

Fig. 5.12 Typical drainage drawing details

Table 5.1 Drainage profile levels table

Position	Design invert level	Interpolated intermediate invert level	'Top of peg' levels	Traveller (m)	Distance from top of peg to top of profile
FMH36	48.750	48.750	51.004	3	↓ 746
+ 5		48.675	51.052	3	↓ 623
+10		48.600	51.060	3	↓ 540
+15		48.526	51.074	3	↓ 452
+20		48.451	51.190	3	↓ 261
+25		48.377	51.202	3	↓ 175
+30		48.302	51.205	3	↓ 97
+35		48.227	51.209	3	↓ 18
+40		48.153	51.242	3	↑ 89
FMH37 (+42.4)	48.117	48.117	51.253	3	↑ 136

1. Set out 1.2 m 'lead' stakes at 5 m intervals, 2 m behind drain centre-line (to face of stake) as Fig. 5.13
2. Drive in 'support' stakes approximately 600 mm behind 'lead' stakes ready to receive profile board
3. Interpolate levels for intermediate positions (manhole invert levels given on drawing) – enter in third column of Table 5.1
4. Level 'lead' stakes, and enter the values in fourth column of Table 5.1
5. A traveller height must be selected – take a random 'top of peg' level and its corresponding 'intermediate invert level', deduct one from the other and round up the result to the next full half-metre or metre increment (e.g.

FMH36 + 20: peg level 51.190 – invert level 48.451 = 2.739 – next increment 3 m)

6. Calculate the figures in final column – note that at FMH 36 + 40 and FMH37 the top of the profile board will be above the top of the peg. Use directional arrows as shown, rather than +/−

7. Measure down (or up) at each stake and construct a profile as illustrated in Fig. 5.13 (97 mm from Table 5.1, final column). The boards should face 'uphill' as operators usually prefer to begin at the lower invert and end at the higher one against the flow direction – information on pipe diameters, traveller height, etc. should be marked on the face of each board so that they can be seen from *below* the excavator, and looking in the direction of dig.

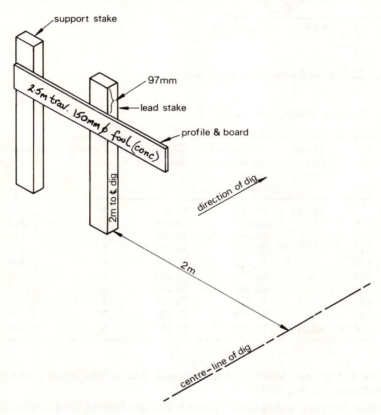

Fig. 5.13 Typical drainage profile

Chapter 6

Setting out: roads

Roads vary in shape and size from tiny private streets in rural settings to the largest inter-city motorways – but the basic site engineering techniques remain exactly the same throughout.

Before preparation of tender documents, the following essential work must be done, often by independent consultants:

(a) land survey
(b) soil investigation
(c) design
(d) provision of vertical and horizontal control on site.

The site engineer is generally not likely to be involved in these pre-contract areas, particularly for larger projects. On smaller works, however, the site engineer may be involved in any or all these activities and particular care should be taken to ensure that both contractor and client are aware that insufficient data had been provided at contract commencement.

The site engineer's normal roads procedure is:

(a) Primary line and level (earthworks, cut and fill)   (Stage 1)
(b) Main drainage line and level   (Drains)
(c) Road base line and level   (Stage 2)
(d) Service ducts line and level   (Ducts)
(e) Kerbing and blacktop line and level   (Stage 3)

At each Stage ((a), (c) and (e) above) it is necessary to set the road centre-line, often at closer centres as the work proceeds.

Circular curve

Most curves used in construction are circular. It is appropriate that several solutions should be given covering various contingencies since curves are frequently used in road design under different site conditions and in inaccessible situations.

There are two stages of calculation:

(a) The 'design' equations
(b) The 'setting-out' equations

Design equations are principally used by the design office at an early stage. In the case of a road curve, radius is dictated by government-issued design standards for a wide range of conditions, and these are published in a table.

The site engineer will not normally be concerned with 'design' parameters, these having been decided and contracted long before his involvement – however, it is often necessary to check through the designer's workings in the preparatory stages, and essential to be conversant with the formulae if minor changes are called for.

Intersection lines

All road design begins with straight lines, and curves are only necessary where smooth changes are required from one straight line to another. The points at which lines intersect are given formal co-ordinates in terms of 'eastings' (E) and 'northings' (N) (the first ones are simply scaled from initial grid lines on the plan, intermediate ones are calculated) and each intersection point is given a reference number, the same reference being subsequently applied to the curve.

'Design' calculations

The various components of the 'design' stage are:

(a) Intersection angle/deflection angle
(b) Tangent length
(c) Curve length

Figure 6.1 shows a typical intersection situation for an estate road. The designer has specified a radius of 40 m, but this does not become relevant until tangent and curve lengths are considered.

The intersection angle is found as follows:

A–B	=	(E) 1067.427 − (E) 1009.927
	=	57.5
B–C	=	(N) 1048.983 − (N) 1047.314
	=	1.669
C–D	=	(N) 1048.983 − (N) 1006.907
	=	42.076
D–E	=	(E) 1077.464 − (E) 1067.427
	=	10.037
ACE	=	$\angle$ ACB + $\angle$ ECD

$$\text{Tan ACB} = \frac{57.500}{1.669} = 34.451\ 767\ 53$$

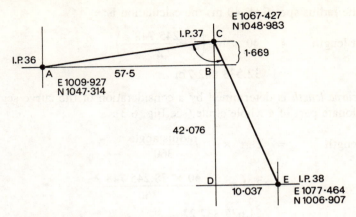

Fig. 6.1 Circular curve intersection

ACB	=	88°20′14″ (88.337 394 54)
Tan ECD	=	$\dfrac{10.037}{42.076} = 0.238\ 544\ 538$
ECD	=	13°25′00″ (13.416 857 24°)

ACE (intersection angle) = 101°45′15″ (101.754 251 8)

The *tangent length* is calculated from:
Tangent length = radius × tan ½ deflection angle
where the deflection angle = 180° − intersection angle, see Fig. 6.2.

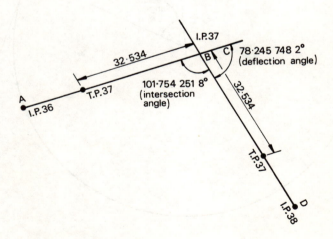

Fig. 6.2 Tangent length

89

For the radius specified (40 m) the calculation is:

$$\text{Tangent length} = 40 \times \tan\left(\frac{78.245\ 748\ 2}{2}\right)$$

$$= 32.533\ 642\ 7\,\text{m}$$

The *curve length* is determined by a consideration of the curve segment as a proportionate part of a whole circle (see Fig. 6.3):

$$\text{Curve length} = 2\,\pi r \times \frac{\text{radius angle}}{360}$$

$$= 2 \times \pi \times \frac{40 \times 78.245\ 748\ 2}{360}$$

$$= 54.625\ 837\ 27$$

(deflection angle = radius angle).

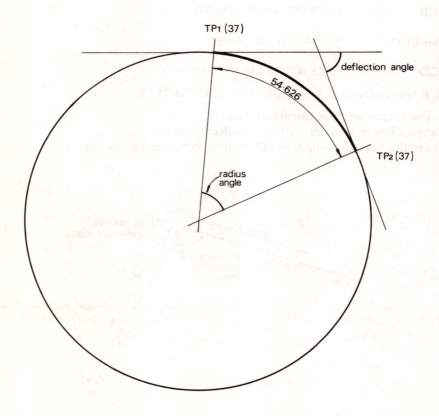

Fig. 6.3 Curve length

'Setting out': calculations and method

There are various methods for setting out circular curves on site, each appropriate in different circumstances.

1. Compass technique
2. Deflection angles (theodolite)
3. Offsets from tangent
4. Quartering

1. Compass technique

Here a measuring tape is used like the radius arm of a compass. Useful for small radii, this involves the site engineer in locating positions of two tangent points and the centre of the circle – the arrangement can then be used by foremen and gangers whenever the need arises (the centre of circle is often referred to as the 'swing point' on site), it has the advantage of great simplicity.

Its limitation is that accuracy suffers when the radius increases to a point where tape corrections become necessary – a physical limitation of the technique.

2. Deflection angles

This is the site engineer's usual method for setting out a horizontal curve, and it relies on the formula:

$$\text{Deflection angle (minutes)} = \frac{1718.9 \times \text{chord length}}{\text{radius}}$$

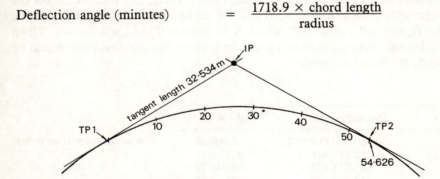

Fig. 6.4 Setting out by deflection angles

Taking the horizontal curve already dealt with in the 'design' section, where the radius was given as 40 m and the curve length calculated as 35.519, the general equation:

$$\text{Deflection angle (minutes)} = \frac{1718.9 \times 10}{40}$$

$$= 429.725' \ (7.162\ 083\ 333°)$$

gives the angle which must be set off for every new chord of 10 m. The tabulation for the curve is:

91

Chords	Decimal degrees	d : m : s	
0	00.000 000 000	00/00/00	(zero on intersection point)
10	7.162 083 333	7/09/43	
20	14.324 166 66	14/19/27	
30	21.486 249 99	21/29/10	
40	28.648 333 33	28/38/54	
50	35.810 416 66	35/48/37	
54.626	39.123 596 41	39/07/25	

– which is based simply on a cumulative progression from 10 m to 50 m, adding 7/09/43 of deflection for each new chord. Measurements are taken from one peg to the next, all round the curve. Pure mathematicians can (and do) argue that the method has minor flaws, but then they do not have to produce consistent results under appalling site conditions. It is sufficient to say that the site engineer will be working to very much finer tolerances using this method than will be achieved by machine operators and tradesmen on site, and will receive no complaints from them.

From Fig. 6.4 it will be noted that with the theodolite set up at TP1, the curve is 'right-handed', i.e. the horizontal circle of the theodolite turns clockwise and the angular figures become larger with each new reading. In a situation where the engineer wishes to set out from TP2, the curve will be left-handed and the angular figures will start at 360° (which is the same as 0°) and will *decrease*. When calculating the curve tabulation, figures for both setting-out directions should be included, for future reference.

Chords	Decimal degrees	d : m : s	
0	00/360.000 000 000	00/00/00	(zero on intersection point)
10	352.837 916 7	352/50/16	
20	345.675 833 4	345/40/33	
30	338.513 750 1	338/30/50	
40	331.351 666 7	331/21/06	
50	324.189 583 4	324/11/23	
54.626	320.876 403 7	320/52/35	

It has been assumed in this second tabulation that the chord lengths will *originate* from positions from TP2, and therefore they will be in different positions from those produced by the first table, although they will of course be along the same line of curvature.

This situation can be corrected, if necessary, by manipulating the chord lengths to accommodate the difference of 4.626 m (54.626 – 54.00) and tabulat-

ing for chords of:
4.626, 14.626, 24.626, etc. to 54.626
and this will ensure precise location of the chord points in exactly the same places
from either direction:

Chords	Decimal degrees	d : m : s
0	00/360 000 000 000	00/00/00
4.626	356.686 820 3	356/41/12
14.626	349.524 737	349/31/29
24.626	342.362 653 7	342/21/46
34.626	335.200 570 4	335/12/02
44.626	328.038 487 1	328/02/19
54.626	320.876 430 7	320/52/35

The setting-out method is straightforward enough. Assuming that the intersection point and tangent lines enclosing the curve are already established, the procedure is as follows (see Fig. 6.4):

1. Set up theodolite at IP37 (using the notation for the earlier example) and sight IP36 (along the same line as TP1), measure out the tangent length to TP1, and locate a peg and nail at that position. Repeat for TP2
2. Move theodolite to TP1 (or TP2, as appropriate) set 00° on IP37 and then set first tabulated horizontal angle, measuring appropriate distance for TP. Locate a peg and nail at that position
3. Set next tabulated horizontal angle, and hook tape on to nail of first curve point, locate peg and nail at second curve point
4. Repeat procedure until curve complete

Obstructions. Should an obstruction be encountered, it should be noted that the theodolite may be relocated on any of the curve points, 0° set and sighted back at the tangent point. The telescope is then pivoted on the *trunnion* axis (without disturbing the horizontal reading) and this provides a new 'artificial' IP. Setting out may then proceed from the top of the tabulation.

3. Circular curve: offsets from tangents
This method is useful when no theodolite is available, but requires the use of an optical square if the offsets become very large. The formula is expressed as:

$$X = R - \sqrt{(R^2 - Y^2)}$$

where X   = offset from tangent
     R   = radius of circle
     Y   = distance along tangent

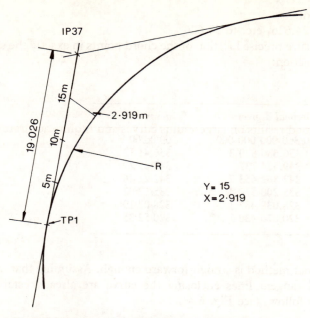

IP37

15m

2·919m

19·026

10m

R

5m

Y= 15
X= 2·919

TP1

Fig. 6.5 Setting out by offsets from tangents

(for Y = 5, the equation would be X = 40 − √(1600−25) = 0.314 m)
From the example used earlier, tabulation would be as follows:

Distance along tangent (Y)	Offset (X)
0	
5	0.314
10	1.270
15	2.919
19.026	4.815

4. Quartering

This is a very quick method of setting out a curve, and is especially useful with road kerbing – indeed, the better kerb-layers use the technique constantly to check alignments and provide intermediate setting-out information. In Fig. 6.6:

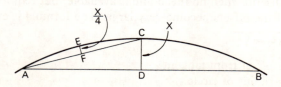

$\frac{X}{4}$

E
C
X
F
A
D
B

Fig. 6.6 Quartering

$$AC \quad \hat{=} \quad \frac{AB}{2}$$

$$\text{offset } EF \quad = \quad \frac{CD}{4}$$

The procedure may be repeated as often as is necessary. Whilst the mathematics is not perfect, and the method relies on various assumptions, the technique will give very good results on large-radius curves and is quite accurate enough for an estate-road kerb-line application where radii are smaller but engineering tolerances often greater.

Vertical curves

Vertical curves are introduced where a relatively sharp change of gradient occurs, and information will usually be given on a longitudinal section drawing.

The designer will have considered:

(a) If a vertical curve is necessary at all
(b) Sight-line requirements
(c) Principal levels and gradients

– and will normally provide the site engineer with calculated principal (tangent) levels and gradients, from (c) above. The site engineer will then calculate tthe intermediate levels, tabulate them and subsequently relate the designer's requirement to site operations.

Because the curvatures are so slight, and the horizontal distances (L) so relatively large, the differences between tangent length, curve length and chord length are negligible and therefore ignored. Similarly, the relative differences at X in Fig. 6.8 are so slight that they can conveniently be forgotten.

Two examples are given for the two basic slope arrangements. Except for some simple geometry to cope with the individual requirements of each case, the same basic formula:

$$X \quad = \quad \frac{a + b}{200L} Y^2$$

is used, for a parabolic curve form.

Example 1

The first example is for a 'summit' curve, and Fig. 6.7 shows part of a typical longitudinal section drawing, as it might be presented to the engineer. Although at an exaggerated scale of 1 : 500 horizontally and 1 : 100 vertically, the curvature involved is still barely perceptible – so the site engineer needs to produce a

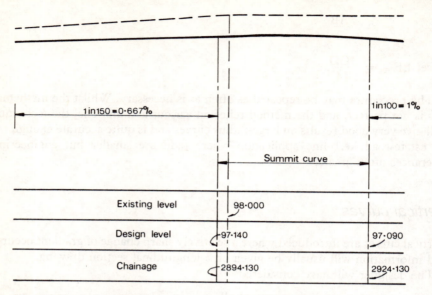

Fig. 6.7 Vertical 'summit' curve: typical design detail

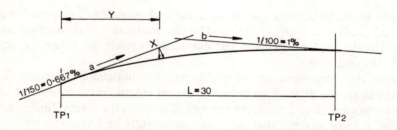

Fig. 6.8 Stylised summit curve

stylised 'sketch' to help examine the relationship more easily. Such a sketch is reproduced as Fig. 6.8 and referred to in the example.

$$X \quad = \quad \frac{a + b}{200L} Y^2$$

$$= \quad \frac{0.667 + 1}{200 \times 30} Y^2$$

$$= \quad 0.000\ 277\ 833\ Y^2$$

Table 6.1 gives AOD values for the centre-line level of the road at 5 m intervals. Adjustments for kerb (etc.) levels must subsequently be made.

Table 6.1 Summit curve table

Distance along curve (Y)	5	10	15	20	25	30
Height of tangent	97.140 +0.033	97.140 +0.066	97.140 +0.100	97.140 +0.133	97.140 +0.167	97.140 +0.200
X (−)	0.007	0.028	0.063	0.111	0.174	0.250
Height of curve (AOD)	97.166	97.178	97.177	97.162	97.133	97.09

Example: for distance 5, 97.140 + 0.033 − 0.007 = 97.166.

Note: height of tangent obtained by $\frac{x}{150}$ = height

i.e. $\frac{5}{150}$ = 0.033, for 5 m from tangent.

Highest and lowest points

From Table 6.1 it can be seen that the lowest point on the curve is at TP2, with a value of 97.09 m AOD, and the highest point falls somewhere between chainage points 10 and 15. The highest point can be calculated from:

$$Y = \frac{aL}{a+b}$$

$$Y = \frac{0.667 \times 30}{1.667}$$

$$Y = 12.004 \text{ m chainage}$$

– the remaining arithmetic is much as before:

'X' for this point is $0.000\,277\,833 \times 12.004^2$	=	0.040 m
Tangent height for 12.004 m	=	0.080
97.140 + 0.080 − 0.040	=	97.180 m AOD

Example 2

This second example is for a 'sag' curve, and Fig. 6.9 shows part of a typical longitudinal section drawing. As with the first example, the scales are exaggerated, but the engineer is still required to sketch out the problem in even more exaggerated form, as in Fig. 6.10.

$$X = \frac{a + b}{200L} Y^2$$

$$= \frac{4.545 + 1.493}{200 \times 60} Y^2$$

$$= 0.000\,503\,167 \; Y^2$$

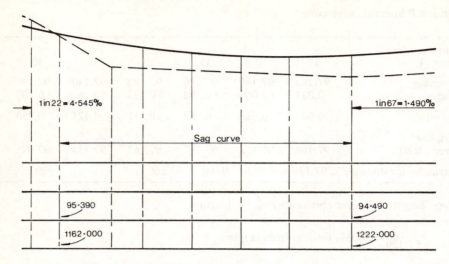

Fig. 6.9 Vertical 'sag' curve: typical design detail

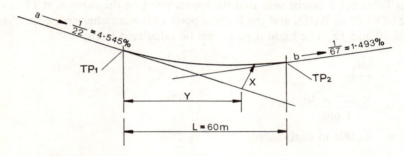

Fig. 6.10 Stylised sag curve

Table 6.2 Sag curve table

Distance along curve (Y)	10	20	30	40	50	60
Height of tangent 95.390 m (−)	0.455	0.909	1.364	1.818	2.273	2.727
X (+)	0.050	0.201	0.453	0.805	1.258	1.811
Height of curve (AOD)	94.985	94.682	94.479	94.377	94.375	94.474

Example: for distance 10, 95.390 − 0.455 + 0.050 = 94.985.

Note: Height of tangent obtained by $\frac{x}{22}$, i.e. $\frac{10}{22}$ = 0.455.

Highest and lowest points

From Fig. 6.10 it can be seen that TP1 is the highest point on the curve, but the *lowest* point is obtained by:

$$Y = \frac{aL}{a+b}$$

$$= \frac{4.545 \times 60}{4.545 + 1.493} = \frac{272.7}{6.038}$$

$$= 45.164 \text{ m from TP1}$$

The actual level at this point is obtained by applying the formula as before:

$$X = 0.000\ 503\ 167\ Y^2$$
$$X = 1.026 \text{ m}$$

Height of tangent $= 45.164 \div 22 = 2.053$

$95.390 - 2.053 + 1.026 \quad = \quad 94.363 \text{ m AOD}$

– and this value is often very useful in determining the position of rainwater gullies, at the lowest point of the curve.

Transition curve

Transition curves are used to smooth out sudden changes in horizontal curvature which might otherwise be dangerous or uncomfortable for the driver. They are normally used in situations where the design speed is above 80 km/hr.

The designer will have prepared much of the curve data before contract commencement using the basic equations:

$$L = \frac{v^3}{0.3\ R} \text{ (assuming a rate of change of radial acceleration of 0.3 m/sec}^3)$$

$$s = \frac{L^2}{24\ R}$$

The site engineer will then prepare setting-out data using the equations:

$$\propto^\circ = \frac{9.55\ y^2}{RL} \text{ (for setting out by theodolite)}$$

$$\text{or } x = \frac{y^3}{6\ RL} \text{ (for setting out by measured offsets)}$$

For the above equations:

L   =   total length of curve in metres and also total length of tangent
v   =   design speed in metres per second
R   =   final radius in metres
s   =   shift
$\propto°$ =   deflection angle in degrees
y   =   a part of L in metres
x   =   offset at 90° from tangent in metres

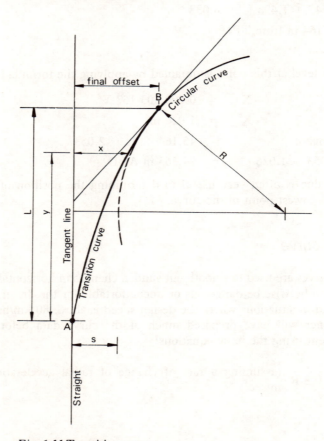

Fig. 6.11 Transition curve

Initial design

Assuming that for the arrangement shown in Fig. 6.11 the final radius (R) is 500 m, and the design speed (v) is 80 km/hr, the curve length (L) is obtained by:

$$L \quad = \quad \frac{v^3}{0.3R}$$

$$= \left(\frac{80 \times 1000}{3600} \right)^3 \times \frac{1}{0.3 \times 500}$$

$$= 73.159 \text{ m}$$

Shift (s) can then be calculated by:

$$s = \frac{L^2}{24 \, R}$$

$$= \frac{73.159^2}{24 \times 500}$$

$$= 0.446 \text{ m}$$

The designer would then use these values in his preparation of contract drawings and curve schedules. At the contract stage the site engineer may then calculate and tabulate data for setting out purposes, using one of the two principal methods.

Method 1 – by deflection angles

The intermediate angles ($\propto°$) are obtained by:

$$\propto° = \frac{9.55 \times y^2}{500 \times 73.159}$$

$$= 0.000261075 \, y^2$$

and the tabulation can be prepared as follows:

Distance along curve (y) metres	Deflection angle d : m : s
(at 'A') 0	00/00/00
10	00/01/34
20	00/06/16
30	00/14/06
40	00/25/04
50	00/39/10
60	00/56/24
70	01/16/45
(at 'B') 73.159	01/23/50

Once the angles have been tabulated there is almost nothing to differentiate this from a circular curve on site. It is quite possible for a novice site engineer to be under the impression that he is setting out a circular curve rather than a transition, if he had had no part in preparation of the table.

Method 2 – by measured offsets

The offsets (x) taken at right angles to the tangent at y metres along the tangent line are obtained by:

$$x = \frac{y^3}{6 \times 500 \times 73.159}$$

$$= 0.000004556 \, y^3$$

Distance along tangent (y) metres		Offset (x) metres
(at 'A')	0	0
	10	0.005
	20	0.036
	30	0.123
	40	0.292
	50	0.569
	60	0.984
	70	1.563
(at 'B')	73.159	1.784

For most site engineering applications, Method 1 will usually be employed since it has the advantage that the engineer can see the curve develop from his distant vantage point and any error will be noticed at once. Method 2 is slower and more cumbersome in practice.

Contract information supplied

The site engineer would not normally be expected to carry out design calculations other than in methods 1 & 2 for setting out. All relevant data must be supplied in the contract documents. Grid co-ordinates for tangent points A and B in Fig. 6.11 should also be provided.

Super-elevation

It is not proposed to deal with the *design* considerations of super-elevation as it would be unusual for a site engineer to become much involved in this. Some brief notes on super-elevation are set out below, but the principal consideration is of the ways in which the site engineer is to express the designer's intentions.

When a vehicle travels at speed around a horizontal curve, centrifugal force is developed and this tends to cause it to slide across the road surface towards the outer edge of the curve. This tendency is corrected by 'banking', and it is theoretically possible to remove all tendency to slide for any given design speed by introducing a specific calculated gradient, although in fact only 40 per cent of the centrifugal effect is generally removed by this means, the remainder being accounted for by the vehicle's tyres.

Super-elevation is fully applied only in conjunction with horizontal circular curves, but it cannot be applied suddenly of course, and is therefore introduced on a 'straight-line' basis over the length of a horizontal transition curve. The curve length must be adequate to allow a straight-line transition gradient of no more than $\frac{1}{100}$ (1 per cent).

To the site engineer the effect is one of a difference in channel design levels (FRLs) and there are virtually no circumstances where a site engineer would be expected to deal with design work of this kind.

Estate roads

It is a matter of common sense that the engineer should not permit a road to be constructed with an 'adverse' fall – i.e. a fall which exaggerates the centrifugal effects of a curve. Given the opportunity, the site engineer should arrange that any road curves not given full design super-elevation consideration should at least have their cross-falls arranged to give what might be termed 'implied super-elevation', i.e. with the cross-fall direction into the centre of the curve.

Centre-lines

Centre-lines have to be reset many times during the course of contract works – usually because the 'centre-line' is only a means to an end and once used (to set off drainage or sub-base profiles, for instance) the original line of pegs becomes redundant. Sometimes the design 'centre-line' is actually in the middle of a carriageway to be excavated to formation level and backfilled with sub-base material. This does not mean that the site engineer must follow slavishly. The site engineer can shift the 'centre-line' data around at will, to suit contract conditions, provided that proper attention is given to compensation for changes in effective radius, such that for instance, a 40 m radius at the centre-line might be 43 m or 37 m at the channel lines, for a 6 m road.

It is often useful, in the final stages of the work, for instance, to set out profiles and other markers direct by theodolite, rather than rely on secondary offsetting techniques.

Sub-base profiles

These are the principal road-building profiles and are used to determine

(a) The road formation
(b) The top of sub-base

To avoid confusion, all site engineering information is based on finished road level (FRL) and adjustments are made in traveller length to allow for:

(a) Design formation level
(b) Design sub-base level

Sub-base profiles are usually set out at no more than 10 m intervals, 1 m behind kerb line, comprising 1.2 m stakes and 600 mm boards. Traveller height is almost invariably 1 m for this configuration, but must suit site conditions.

Setting-out sub-base profiles may only begin when major earthworks are complete, and only the final (say) 200 mm of material is to be removed to expose the formation level. Clearly, it is necessary to control the primary 'major earthworks' carefully if this secondary operation is not to be hampered by the necessity to remove excessive amounts of material, or, perhaps worse, to have to import extra sub-base material just to fill an over-excavated section. The expression 'over-dug' is used on site when this situation arises.

Even in a 'fill' area where subsoil is imported and compacted, there is always a surplus covering which is cut away at the 'secondary' stage to reveal the formation level.

The final cut to formation level is carried out quickly and carefully, and ideally the sub-base material (which compares with the concrete strip foundation under a house) is laid and rolled immediately. Any serious soft spots will be evident as the sub-base is rolled; either the material will not compact properly or perhaps mud will be seen oozing through to the surface. Soft spots are dealt with by immediate excavation and filling with high-quality sub-base material.

Binders

Occasionally a binding membrane is introduced at formation level – a very strong man-made fibre sheet is available which binds to the formation layer and helps to contain the sub-base material, resisting to some extent the tendency for individual groups of stones to punch deep into the soil. Bundles of heather were once used, and they filled the purpose very well. In some instances this material might be more readily available.

Sub-base materials

Many different materials are available, and it is not appropriate to enter into a discourse on sub-base specification in this book, as this is not strictly the premise of the site engineer. Any sub-base material must, however, conform to some basic requirements to be of use on site and these are:

(a) It should be 'clean' and free from obvious impurities, foreign bodies, organic matter, etc.
(b) It should be free of 'fines' (i.e. clay, silt, etc.) to the naked eye
(c) It should be reasonably well graded, between (say) 75 mm and 10 mm

(d) It should be sharp, angular material, capable of binding and compaction. Smooth, rounded stones will probably not bind successfully and will be difficult to compact

If the site engineer suspects that the sub-base material is not satisfactory he should refer the matter to the consultants immediately.

Single profiles

When setting out sub-base profiles, account must be taken of road cross-falls. Figure 6.12 shows a 9.6 m carriageway with a single cross-fall (one side of a typical dual-carriageway arrangement). Channel levels A and B may be supplied with the contract drawings for each section or may have to be calculated by simple geometry, but the levels A and B (top of profile) are calculated as follows:

A1 $\frac{1}{80} \times 1$ m (horizontal offset) $= 0.012$ m

$103.259 + 0.012$ m $+ 1$ m (traveller height) $= 104.271$

B1 $\frac{1}{80} \times 1$ m (horizontal offset) $= 0.012$ m

$103.139 - 0.012 + 1$ m (traveller height) $= 104.127$

Note: All levels are based on the finished road level (FRL) in each case.

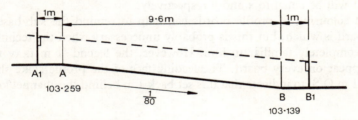

Fig. 6.12 Single profiles

Twin profiles

A slightly more complex situation arises when a cambered road is involved, and this is dealt with by use of 'twin' profiles – where two boards are nailed to each stake, and the bottom board on one side is sighted across to the top board on the other, for only half of the road construction. Figure 6.13 shows a typical estate road of 6 m width. Channel levels A and B would normally be the same, and set out directly, with the centre-line (crown) level arrived at, on site, by indirect means.

105

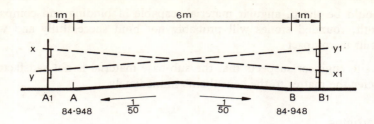

Fig. 6.13 Twin profiles

The heights of profile boards x and y at stake position A, are calculated as follows:

For x, $\frac{1}{50} \times 7$ m (carriageway + 1 m offset) $\quad = \quad$ 0.140 m

84.948 + 0.140 + 1 m (traveller height) $\quad = \quad$ 86.088
$\qquad\qquad\qquad\qquad\qquad\qquad\qquad\qquad\qquad$ (height of x)

For y, $\frac{1}{50} \times 7$ m (carriage + 1 m offset) $\quad = \quad$ 0.140 m

84.948 − 0.140 + 1 m (traveller height) $\quad = \quad$ 85.808
$\qquad\qquad\qquad\qquad\qquad\qquad\qquad\qquad\qquad$ (height of y)

x1 and y1 will be equal to x and y respectively.

Special colouring of profile boards can help to remind the sub-base ganger which board is which, but this is probably unnecessary when the technique has become completely familiar to him. However, the legend '1 m TRAV to FRL' should appear on every board. The leading face of the profile stake should be marked '1 m O/S' to indicate that it is set back 1 m behind the channel/kerb line.

Setting out

Profile stakes are usually set out from the centre-line pegs, themselves placed at the correct centres in readiness, using a hand-held optical site square and tape. Offsetting from centre-line pegs by this technique can be difficult initially owing to the relatively small size of the pegs. It is often necessary to crouch down momentarily to obtain the sight-line, then fix the unaided eye on some detail beyond the stake position – then to direct the chainman to line in with the observed detail. A final check can then be made optically. With practice, the engineer can become very adept at this procedure, and make it look deceptively easy.

A complication arises when setting out 'square' to a curve – at right angles to the tangent. This is illustrated in Fig. 6.14 and is actually very simple, as the hand-held site square is simply moved back or forward until the pegs A and C

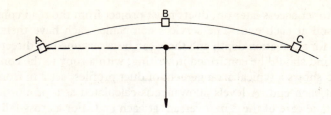

Fig. 6.14 Setting out from centre-line pegs

apparently intersect. The engineer will probably lose contact with B during this operation, so the measurement must be taken separately.

Service ducts

It is invariably necessary to lay services across a roadway, and the usual ones are:

(a) Mains water pipes
(b) Electricity cables
(c) Gas pipes
(d) Telephone cables

Some more unusual requirements are for:

(e) Television cableways
(f) Computer links
(g) Traffic signal wiring
(h) RT repeater stations
(i) Oil pipework

– and there are of course many others.

To the site engineer the purpose for which the duct is to be used is almost irrelevant, but some general requirements are common to all:

1. The duct should be laid to a slight fall so that any water entering does not lay
2. Excavation depth should be sufficient to ensure that the duct will not be damaged when the roadway is in use
3. A draw-string, toggle set and some form of identification should be installed with the duct

Duct profiles

Duct-laying is often carried out after most of the sub-base material has been placed in position, and therefore it is common practice for the site engineer to use 'relative' levels for ducts, using data from nearby sub-base profiles and from the top of the sub-base level.

To allow proper access later on, ducts must project from the road construction into natural soil at each end. The services companies each have their separate requirements for excavation depth and these can be ascertained by direct enquiry. Data of this kind should be confirmed in writing, with a copy to the consultants.

Figure 6.15 shows a typical arrangement of duct profiles, set 2 m from the end of the duct at each end. A levels allowance is calculated as a proportion of the cross-fall, to take care of the 2 m 'oversail' at each end. For a cross-fall of 1 : 50 the allowance would be +40 mm at A, and −40 mm at B.

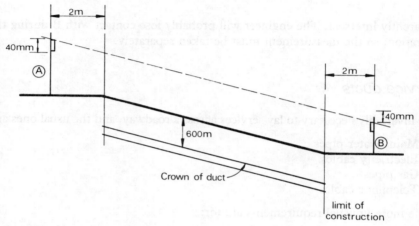

Fig. 6.15 Ducts

Before setting out duct positions and depths on site, the engineer should establish:

(a) Minimum permissible 'cover' (FRL to crown of duct) for specific types of duct, from the consultants
(b) Specific type of duct(s) to be used in which locations
(c) Depth requirements, and any other stipulations, laid down by the services companies
(d) Any other ducting requirements, apart from those normally anticipated (some site engineers will authorise 'spare' ducts at intervals in anticipation of a future problem)

Warning

If ducts are not properly installed, or cannot be easily found, the services companies will invariably excavate a 'service trench' across a newly finished road, cutting through the wearing course, base course and sub-base – disturbing kerbs and drainage in the process.

Reinstatements are rarely satisfactory, always expensive, and problems of this kind can prove very embarrassing to both the site engineer and his employer.

Finding buried ducts

If he has done his job properly, the site engineer will be able to identify duct positions by taking very simple taped measurements from fixed points, and 'squaring' across the road. Probably the best system is to relate all duct positions to the drains layout, using manholes as fixed points and measuring 'from FMH 16, 11.7 m towards FMH 17' – it is often possible to deal with the problem by telephone provided the other party has a basic drains layout, which should not be difficult to arrange.

It is very important to ensure that each of the services organisations only uses ducts allocated to it, for obvious reasons, and the ideal way to ensure this is to supply the foreman with a photocopied 1 : 500 drawing showing only the relevant duct crossing marked, with site measurements carefully underlined. This makes more work for the engineer, but may be considered worth while.

A 'toggle' comprising a 1.5 m length of 100 × 50 mm timber, attached to a 19 mm nylon (rope) draw-string passing through the duct, should be fairly easy to find, and can either be left sticking out of the ground or be left buried below the surface across the line of the main service route, in which case it will eventually be discovered by accident even if the plan information is lost or ignored.

Road pins

At this stage there is usually no point in resetting centre-line positions used earlier, as this would probably mean driving a further set of road pins into the sub-base, and the channel-line pins would only be to 'optical site square' accuracy.

By manipulation of the horizontal curves data, adding to or subtracting from the design radius to allow for 'outside' or 'inside' curves, it is possible for road pins to be set out directly on to channel lines or, alternatively, back of kerb lines to become direct 'kerb pins', first on one side of the carriageway and then on the other. Spacing is usually no more than 10 m, and often 5 m, centres.

Road pin levels

Levels are taken on the tops of each of the pins when a whole section has been completed. Tapes or 'jubilee' clips are fixed in position usually 100 mm above FRL, and this applies both to kerbing work (where a 100 mm face is correct) and to blacktop without kerbing, where the 100 mm-above FRL measurement is used by the blacktop ganger to check his machine's laying height at each 5 m or 10 m section interval. Where kerbs are already in position, these provide a constant guide.

Dips

To assess the required thicknesses of base and wearing courses prior to laying, measurements are taken below a stringline stretched taut across the roadway at (typically) 2m centres, along the string, at each road pin section of 5 m or 10 m. These measurements are called 'dips' on site and can form the basis of a contract claim. If the roadway is kerbed prior to blacktopping, 'dips' may be taken below a stringline pulled taut over the tops of kerbs, at the same intervals as before. In this event, yellow road-chalk markings are used to denote the section chainage intervals.

Chapter 7

Mass earth-moving

The most common reason for mass earth-moving is the construction of roads, and this chapter will refer mainly to highway construction – the main principles are just the same for many other applications, however, and can be applied equally well to a terraced housing site or a deep construction pit for oil-platform manufacture.

First things first

Mass excavation is one area where the site engineer if often expected to accept much more responsibility than would be considered normal for other site operations, and he must therefore ensure that he asks the right questions (and has them answered) before allowing the work to proceed. Before the excavators and lorries arrive on site, particularly for a large project, the site engineer should satisfy himself on a number of points:

1. The earth-moving is necessary
2. All possible savings and economies have been made
3. There is a suitable and adequate tip
4. Road-closure and vehicle-licensing arrangements are agreed with the authorities, in writing
5. Excavator capacity is matched by lorry capacity, bearing in mind actual turn-round times (recorded in trials)
6. The plant supplier has made up-to-date arrangements for fuel, tyres, servicing and repairs to all vehicles involved
7. The weather is likely to be good enough, and a twice-daily forecast is arranged
8. All insurances and driving licences are valid (a written statement from the plant supplier on driving licences will be adequate)
9. The consultants are satisfied with the quality of the soil to be used for filling
10. All work can be finished by the programme date, subject only to weather
11. All necessary setting-out is done

Setting out

The centre-line must first be set out, as a series of straight and curved sections, at intervals of generally 30 m. Pegs at 10 m intervals may be preferred for some curves. Pegs marking tangent points should be coloured (say) red to clearly distinguish them, and chainages should be marked on every peg.

The horizontal position of the centre-line will not alter throughout the project, but the surface on which the first centre-line pegs are placed will either be substantially raised or lowered, resulting in obliteration of any pegs.

The next operation is to position angled profile boards to indicate the exact point of commencement of the new slope, and the slope angle. Figure 7.1 shows the two basic types, both are generally referred to as 'batter rails' on site, although a more correct term would be 'primary earth-moving profiles'. It should be noted that 'cutting' profiles are set back 1 m from the top of batter, but the profile board is directly in line with the slope. 'Embankment' profiles are set exactly at the bottom of the intended slope, but the profile board is set *above* the line of slope and a 'traveller' must be used.

Alternative styles of profiling may be adopted – there is no reason to adhere to those illustrated if they do not fulfil a particular requirement. This is, however,

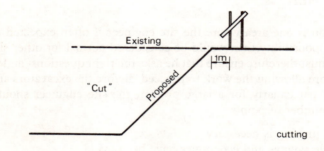

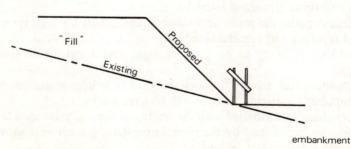

Fig. 7.1 Batter rails

one area where the matter should be agreed with the plant contractor at the outset, and to avoid any subsequent confusion each machine operator should be supplied with a copy of the agreed detail.

Before any profile stakes can be driven in, some quite exacting survey/calculation work must be undertaken – using the centre-line pegs as datum points.

To the site engineer who has never done it, or thought it through, the problem of locating the top or bottom of slope appears deceptively simple. It is in fact a relatively complex exercise in geometry, involving several site and office operations.

Determination of side-widths

For site engineering purposes, side-widths can be obtained by:

(a) 'Graphical' method
(b) Office calculation
(c) 'Direct' method

Graphical method

This requires a detailed survey of cross-section levels at centre-line points, and the data obtained plotted on special cross-section extracts from the main contract drawings. The principal excavation lines are drawn in and the side-slope inscribed using a template (set to the predetermined slope angle notified by the consultants). Side-widths may be scaled from the drawings.

Office calculation

The formulae given in Fig. 7.2 may be used to calculate side-widths (more accurate than scale measurement), but some levelling must still be carried out before calculation.

Direct method

This method has the very distinct advantage that there is no necessity for office work between initial levelling and final setting-out stages, and is one of those techniques which look complex in print but which are very simple to operate – an experienced site engineer will quickly acquire a 'feel' for the procedure, as in common with so many site engineering techniques this is one of successive approximation to a point where the error becomes insignificant.

The purpose is to find the position of B relative to G. The method is exactly the same for embankments (Fig. 7.3a) and cuttings (Fig. 7.3b).

1. Take a level at G
2. Deduct level G from level A, leaving V

113

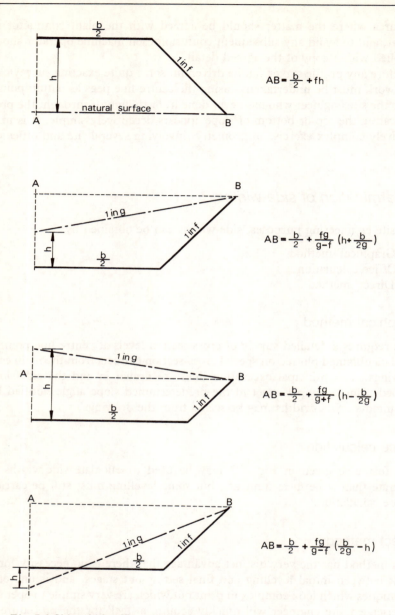

$$AB = \frac{b}{2} + fh$$

$$AB = \frac{b}{2} + \frac{fg}{g-f}\left(h + \frac{b}{2g}\right)$$

$$AB = \frac{b}{2} + \frac{fg}{g+f}\left(h - \frac{b}{2g}\right)$$

$$AB = \frac{b}{2} + \frac{fg}{g-f}\left(\frac{b}{2g} - h\right)$$

Fig. 7.2 Side-width office calculations

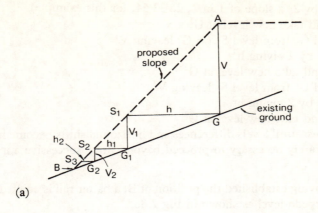

(a)

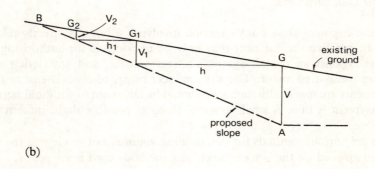

(b)

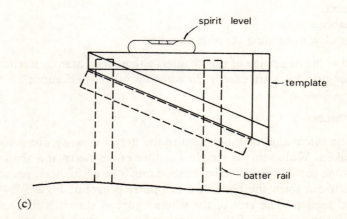

Fig. 7.3 (a) 'Direct' embankments; (b) 'direct' cuttings; (c) template for batter rails

115

3. Multiply V by 2 (a slope of 1 in 2, 26/33/54, for this example)
4. Set out h and take new level at G^1
5. Deduct level G^1 from level S^1 (= G) leaving V^1
6. Multiply V^1 by 2 giving h^1
7. Set out h^1 and take new level at G^2
8. Deduct level G^2 from level S^2 leaving V^2
9. Multiply V^2 by 2 giving h^2
10. Set out h^2 and take new level at G^3 . . .
11. Repeat process until levels differences and horizontal shifts become insignificant – it is rarely necessary to proceed beyond three successive approximations.

In both cases, having established the position of B, a batter rail is erected using a template and torpedo level as shown in Fig 7.3c.

Volume calculations

To the site engineer almost any question involving volume is directly related to money. In practice the engineer may only be responsible for authorising 'stage payments' against an overall agreed contract volume, and any deficit will be settled on the last payment. The site engineer must, of course, ensure that no overpayments are made, although care should be taken not to withhold noticeably where payment is due. A small 'reserve' to cover possible slight inaccuracies is permissible.

There are various methods for determining volume, but in view of the special constraints placed on the site engineer, any methods used must be:

(a) Quick to give results
(b) Simple to understand
(c) Repeatable
(d) Reasonably accurate
(e) Capable of withstanding scrutiny in court

As stated at the beginning of this chapter, although constant reference is made to roads, the principles are exactly the same for other applications.

Areas of sections

Most volume calculation methods require the use of cross-sections from which areas are taken. Whilst this is simply a matter of geometry and arithmetic, an example of one particularly effective method might be useful here, particularly as the approach can form the basis of a computer program. Figure 7.4 shows a typical 'cut' section. The area of the whole figure as shown is first calculated, multiplied by the average of the vertical ordinates. Areas 1, 2 and 3 are then calculated and deducted.

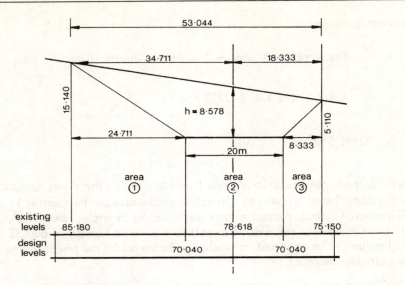

Fig. 7.4 Typical 'cut' section

Total area $= 53.044 \times \dfrac{85.180 + 75.150}{2}$

$\qquad\qquad = 4252.272 \ \text{m}^2$

Area 1 $= \dfrac{24.711 \times 85.180 + 70.040}{2}$

$\qquad\quad = 1917.8207 \ \text{m}^2$

Area 2 $= 20 \times 70.040$

$\qquad\quad = 1400.8 \ \text{m}^2$

Area 3 $= 8.333 \times \dfrac{70.040 + 75.150}{2}$

$\qquad\quad = 604.934 \ \text{m}^2$

Excavated area $=$ total area $-$ (Area 1 + Area 2 + Area 3)
$\qquad\qquad\quad\ = 4252.273 - (1917.8207 + 1400.8 + 604.934)$
$\qquad\qquad\quad\ = 328.7183 \ \text{m}^2$

Sections *(chainages)*	(m)	180	210	240	270	300	330	360	390
Area (exc)	(m²)	328.7	421.4	329.3	284.6	585.9	594.6	496.7	412.4

Note: These are *gross* areas, and any earlier volume payments must be deducted from the final volume.

End-area formula

$$V = \frac{d}{2} \text{ (first area + last area + 2 × sum of intermediates)}$$

$$V = \frac{30}{2} (328.7 + 412.4 + 2 (2712.5))$$

$$V = 92491.5 \text{ m}^3$$

– this is the total volume extracted so far from the site, but there may have already been payments based on (say) 81 450 m³, so the residue for payment is 11 491.5 m³. The overall agreed contract volume might be, for example, 124 945 m³, and it would be of interest to the contractor at this stage to let him know that 32 453.5 m³ still require to be extracted. It would also be useful for the project manager to know that (for instance):

(a) 74.025 per cent of the contract volume has been excavated to date
(b) 9.197 per cent of the overall contract volume was excavated last week

The site engineer can therefore provide various types of information from one simple calculation.

Simpson's rule

Another method of volume calculation using cross-section is available, called 'Simpson's rule for volumes' where

$$V = \frac{d}{3} \left[\frac{\text{first}}{\text{area}} + \frac{\text{last}}{\text{area}} + 4 \times \frac{\text{sum of even}}{\text{intermediate areas}} + 2 \times \frac{\text{sum of odd}}{\text{intermediate areas}} \right]$$

– and in this form an odd number of sections are required.

Volumes from spot heights

Volumes may be computed from 'grids' of levels taken before and after excavation. The site work required is tedious and exacting, so the method is generally restricted to very small areas – setting out and levelling a large grid is a very expensive and time-consuming operation, and the subsequent office work somewhat exhausting. Each grid square is considered separately, and clearly the smaller the grid centres the greater the accuracy. For more accuracy, triangles may be considered. It will be noticed in Fig 7.5 that some grid points form corners in only one square, some in two and others in four.

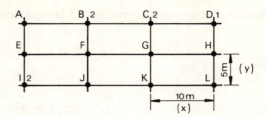

Fig. 7.5 Volumes from spot heights

It is possible to simplify the arithmetic by grouping the grid points into 1's, 2's and 4's, and a complete algebraic expression can be built up as follows:

$$V = \frac{xy}{n} \left[(A + D + I + L) + 2(B + C + E + H + J + K) + 4(F + G) \right]$$

where n = number of ordinates

Example

A tank with a sloping bottom has a length of 30 m and a width of 10 m (example based on Fig 7.5, x = 10 m, y = 5 m). Levels and other details are as follows:

Grid point	Existing level	Design level.	Depth
A	56.948	52.50	4.448
B	56.823	52.90	3.923
C	55.954	52.30	2.654
D	56.103	53.70	2.403
E	55.997	52.50	3.497
F	56.018	52.90	3.118
G	56.724	53.30	3.424
H	56.548	53.70	2.848
I	56.732	52.50	4.232
J	55.922	52.90	3.022
K	57.180	53.30	3.880
L	56.995	53.70	3.295

$$V = \frac{10 \times 5}{12} \times \left[(4.448 + 2.403 + 4.232 + 3.295) \right.$$
$$+ 2(3.923 + 2.654 + 3.497 + 2.840 + 3.022 + 3.880)$$
$$\left. + 4(3.118 + 3.424) \right]$$
$$V = 1023.021 \text{ m}^3$$

A gross error check

As a general check, the average of all the depths may be taken without regard to frequency of appearance, and any gross error should become immediately apparent:

4.448
3.923
2.654
2.403
3.497
3.118
3.424
2.848
4.232
3.022
3.880
3.295
————
40.744

$$\frac{40.744}{12} = 3.395$$

$3.395 \times 30 \times 10 = 1018.6 \text{ m}^3$ (compares with 1023.021 m^3)

– this method has no validity other than as a rough check for gross error.

The mass-haul diagram

The mass-haul diagram is a useful site management tool which can assist in:

(a) Identification of surpluses or deficiencies of suitable fill material on site
(b) Prevention of
    (i) double handling
    (ii) uneconomic 'overhaul'
    (iii) adverse effects of other works
(c) Selection of correct types and numbers of excavators and lorries

To produce a mass-haul diagram, various items of information must be available:

(a) A longitudinal section through the site
(b) Areas of cross-sections at regular intervals (30 m) and consequently cut/fill volumes between sections
(c) Bulking/shrinkage factors for the materials to be excavated

A longitudinal section is shown in Fig. 7.6, a related mass-haul diagram in Fig. 7.7 and a mass-haul table (Table 7.1).

The mass-haul diagram is simply a graph, plotting chainage against cumulative bulked volume values, extracted from Table 7.1. Where the curve is rising (above or below the horizontal axis), excavation is taking place. Where the curve is falling, filling is taking place.

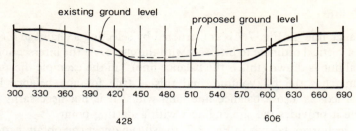

Fig. 7.6 Longitudinal section

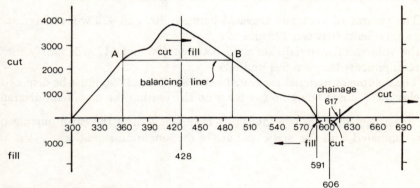

Fig. 7.7 Mass-haul diagram

Table 7.1 Mass-haul table

Chainage	Volume (m³)		Bulking factor	Bulked volume	Cumulative bulked volume
	Cut	Fill			
300	—	—			
330	824	—	1.33	+1096	+1096
360	943	—	1.33	+1254	+2350
390	325	—	1.33	+ 432	+2782
420	784	—	1.33	+1043	+3825
450	—	555	—	− 555	+3270
480	—	647	—	− 647	+2623
510	—	798	—	− 798	+1825
540	—	835	—	− 835	+ 990
570	—	810	—	− 810	+ 180
600	—	596	—	− 596	− 416
630	653	—	1.20	+ 784	+ 368
660	532	—	1.20	+ 638	+1006
690	621	—	1.20	+ 745	+1751

Notes:
1. Bulking factors 1.33 and 1.20 obtained from consultants for different types of soil.
2. General assumption that excavated material suitable as fill.
3. For the section 1300–1690 there is a surplus of 1751 m³ to be passed on.

121

Balance line

The 'balance line' in Fig. 7.7 is constructed by selecting a chainage (360 m) and drawing a vertical ordinate to intersect with the curve. At the intersection point, a horizontal line is drawn (this is the balance line) – in the example the cut volume exactly matches the fill volume, and this is true for any balance line. The horizontal axis is also a balance line, and whilst it is not necessarily the most useful one it provides the site engineer with a starting-point.

Using the horizontal axis in Fig. 7.7, the site engineer can obtain various items of valuable information from this mass-haul diagram:

(a) The volume of material excavated between 300 and 428 will satisfy the fill requirements between 428 and 591
(b) the volume of material excavated between 606 and 617 will satisfy the fill requirements between 591 and 606
(c) The volume of material excavated between 617 and 690 must be disposed of elsewhere, there being no use for it on site (within the mass-haul diagram).

The mass-haul diagram is a 'dynamic' representation of the earth-moving process, compared with the 'static' picture of the longitudinal sections.

Chapter 8

Tests and investigations

Concrete

In practice the site engineer will be involved in only two kinds of concrete test on a day-to-day basis, and will have neither the time nor the expertise to become more deeply involved in analytical techniques. Large projects will sometimes have a resident concrete technician who will continuously inspect, grade and analyse materials and mixes, and smaller (but substantial) jobs may employ a visiting technical consultant.

Ready-mixed concrete

The usual reason for using ready-mixed concrete is that it is convenient; there is no waste, no materials stockpiling and no extra labour requirement. The companies involved in this business are usually very co-operative, and will make all kinds of special arrangements to suit the site's requirements. The usual reason for *not* using ready-mixed concrete is that site-mixing is cheaper – and as a 'rule-of-thumb' this is often true if the quantity involved is either very large (and over a long time period) or very small.

Most ready-mixed concrete firms will supply on a 'collect' basis, where the contractor sends his own tipper lorry to the batching plant and collects the quantity required. This is usually considerably cheaper than having concrete delivered to site.

One of the great advantages to the contractor of using ready-mixed concrete is that the material can be specified by *strength* (as required by the design consultants) and the ready-mixed concrete firm will *guarantee* the final strength or take full financial responsibility for the consequences. The site engineer should not, however, be lulled into a false sense of security by these assurances, and should check:

(a) Each incoming load's delivery note for the correct specification, as ordered
(b) The 'slump' of random loads
(c) The crushing strength of random loads (at least one per critical structural element)

Detailed requirements for ready mixed concrete are laid down in BS 5328.

Delivery notes

Each load of concrete delivered to site will be accompanied by a 'delivery note' and carbon copy, which is returned with the empty lorry. A signature is required at the bottom of the sheet, and it is simple to give instructions when telephoning the order that delivery notes are to be signed only by one person (the site engineer).

It is then a simple matter to read the mix specification and other details before signing. After handing the carbon copy back to the driver, the location and purpose of the pour may be written on the back of the top copy, before it is passed on to the accounts department for filing. By this simple means, an immediate, if cursory, check is made on all incoming loads of concrete, and an impressive array of information can be produced at a later date if necessary.

Slump test

This is a quick check on 'workability' and is unrefined because various highly changeable factors affect it. They are:

(a) Water/cement ratio
(b) Cement/aggregate ratio
(c) Aggregate porosity, texture and shape
(d) Effective surface area of aggregate

Concrete must have 'workability' so that it may be spread easily and compact readily, but effectively; too much water in the mix will result in concrete weakened by:

(a) Globules of free water trapped in the mass
(b) Aggregate washed clean of cement particles
(c) Aggregate sinking to the bottom of the mass, leaving over-rich material at the top

The slump test provides a rapid workability reading for immediate comparison of one batch with another. The apparatus consists of:

(a) 600 mm × 600 mm × 6 mm steel baseplate
(b) 300 mm high slump cone, 100 mm dia. at top, 200 mm dia. at bottom
(c) 16 mm dia., 600 mm long bullet-nosed tamping rod
(d) Small filling shovel
(e) Tape-measure

The cone is fitted with handles to facilitate lifting, and small foot-rests so that it can be held in position while being filled.

The cone is filled with test concrete in four distinct layers, each layer being tamped with the rod twenty-five times. When the cone is filled, the top is

124

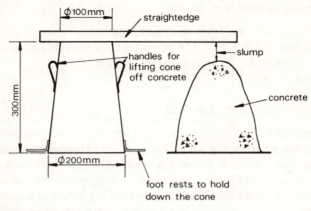

Fig. 8.1 Slump test

smoothed off using the tamping rod. The cone is then lifted off the concrete in one smooth vertical movement, and immediately placed beside the standing concrete. Height reduction is then measured, by holding the tamping rod across the top of the steel cone and measuring down from its underside to the top of the concrete. The height reduction is termed the 'slump' and should be recorded, together with details of the delivery note number, and the exact purpose and location of the batch. Figure 8.1 shows the final stage of a typical test.

Crushing strength

The most reliable method of establishing the strength of something is to break it under controlled conditions, and this is done with concrete – usually in the form of 150 mm × 150 mm × 150 mm cubes moulded from samples taken at the point where the material is to be used. The cubes are subsequently taken to a laboratory where they are placed one at a time in a test apparatus, and loaded gradually until they fail. The failure load is recorded by an instrument on the machine, and converted by the operator into a stress in N/mm^2. This figure is then noted against the cube's number and description in a test report sheet, and subsequently forwarded to the site engineer.

BS 1881 gives full particulars of sampling and testing procedures, and the following text contains details taken from this source.

Making the cubes

Test cubes must be prepared to a standard pattern if the failure-load results are to be realistic, representative and comparable with those from other cubes. Three test cubes should be prepared for every sample, one to be crushed at seven days, another at twenty-eight days and the third as a 'control' cube if the twenty-eight-day test proves inadequate.

The equipment needed is as follows:

(a) 600 mm × 600 mm × 6 mm steel baseplate
(b) Three 150 mm × 150 mm × 150 mm steel cube moulds and suitable spanner
(c) 25 mm × 25 mm × 375 mm steel ram, weighing 2 kg
(d) Small filling shovel
(e) Sacks (damp)
(f) A water tank, preferably in a shed

The spanner size for the nuts and bolts holding cube moulds together is usually the same as for a car spark-plug – and this has both advantages and disadvantages. It is suggested that a suitable spanner is chained to the water tank.

The water tank should have a submerged shelf, preferably of mesh, so that water can circulate freely. It should be covered when not it use, preferably be sited in a shed reserved solely for the purpose and maintained at 20° C ± 2° C.

Moulds are filled with concrete in four distinct layers, each layer being rammed a minimum of thirty-five times using the standard 2 kg ram. When full, the top of the concrete is smoothed off using the ram. The moulds and their contents should be stored under a damp sack for twenty-four hours at not less than 15° C. Then the concrete cubes may be released – great care is necessary to avoid damage – and reference numbers may be marked clearly on the *top* surfaces. All three cubes should then be placed together in the water tank, in a clearly identifiable group, and not only their descriptions and reference numbers recorded *but their positions in the tank*. The moulds should then be cleaned, oiled and reassembled for re-use.

Validity of tests

The crushing strength test results obtained are an indication of the strength and quality of the concrete tested, but do not necessarily indicate the actual strength of the concrete when it is in place, since the degree of compaction may differ considerably between site application and sample. It is therefore worth while to ensure that site concrete and sample are compacted in as near-identical a manner as possible, even to the extent of using a poker vibrator on the sample if one is used on site.

If the samples fail

In the event of tests showing, perhaps five or six weeks after pouring, that the concrete samples are inadequate, the concrete suppliers must be immediately informed in writing, and provided with a copy of the test report. They should also be informed of the exact location of the alleged defective material, and the use to which it was put.

Many concrete suppliers take sample cubes of their own whenever a 'strength' mix is ordered (as distinct from a mix specified by cement: aggregate ratios) and may be able to produce a cube result which gives a higher value. Alternatively, they may wish to drill out a circular section 'core sample' from the actual structural component involved, and subject to the consultants' agreement on a suitable location, one or more samples may be removed for crushing.

A cylindrical core sample is crushed in the same test apparatus as a 150 mm cube, but is supported in a 'V' block to stop it rolling around. Conversion of the actual load at failure to an expression in units of N/mm^2 is slightly more complicated than with a standard cube, because a cylindrical sample loaded on its curved surface tends to fail in a single verticle crack (a tensile failure mode) – but nevertheless a reliable result is obtainable.

If it is shown that the concrete used in the work is under strength, and the consultants are unable to accept it as permissible in the isolated case, the structural element must be broken up and replaced with new materials. In some instances this can be enormously expensive.

Drains tests

Although in some highly specialised situations drain pipework may be designed to be 'porous', this is usually to allow groundwater or surface water to *enter* the pipe system. In general terms, however, drains should not leak at all. If they do, and are carrying solids in the form of faeces and/or kitchen waste, two things are likely to happen:

1. There will be insufficient water to lubricate the passage of the solids and a blockage will result
2. The escaping water will, over the years, leach away particles of bearing soil beneath any adjacent foundations and lead to settlement

There are two basic types of drains test:

1. Hydraulic or 'water' tests (pressurised, unpressurised)
2. Pneumatic or 'air' tests (low pressure, high pressure)

Hydraulic tests

Unpressurised water
An unpressurised water test involves little more than putting an expanding stopper in the lower end of a section under test, and filling from the upper end with water. When full, a slight drop in level is to be expected in the first minute or so of the test, owing to water being absorbed by the slightly porous pipes (if clayware) and slight lossess into the joints. The pipe may then be topped up, and should remain absolutely full and steady for five minutes Fig. 8.2 shows a situation in which this type of test is particularly useful, where it is perhaps not practical or economic to air-test the whole SVP (Soil and Vent Pipe). The level of water will be the same in both the SVP and the clear plastic testing tube, which is

simply held steady against the manhole rim for the duration of the test. A slight fluctuation in level might be noted, both upward and downward, due to slight atmospheric pressure changes in the SVP, brought about by wind passing over the open top of the pipe.

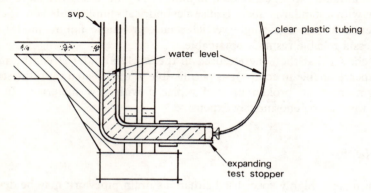

Fig. 8.2 Unpressurised water test

Pressurised water

A pressurised water test involves some rather more complicated equipment, including a long glass tube marked with graduations, and is based on the principle that a 1 m head of water will stress the drain system sufficiently to show up any leaks undetected in an unpressurised test. This technique is of doubtful value to a busy site engineer since it is rather time-consuming to set up and conduct.

Pneumatic tests

Low pressure

This is probably the most common form of drains test, and is much more severe than an unpressurised water test. It requires that a section of drain be plugged at both ends with expanding stoppers, one of which has a nipple at the centre through which air may pass into the section under test. A rubber tube is connected to the nipple, and air is pumped in to the drain run by means of a small hand-pump, often in the form of a rubber bulb, fitted with a non-return valve. The tube is then pinched to retain the air pressure, the pump removed and the tube connected to a half-filled 'U' gauge.

A slightly more sophisticated arrangement is shown in Fig. 8.3, where the pump is permanently connected into the test apparatus, saving much fumbling with rubber pipes and nipples. This arrangement also has the advantage that any required 'start' reading can be easily achieved on the gauge without guesswork.

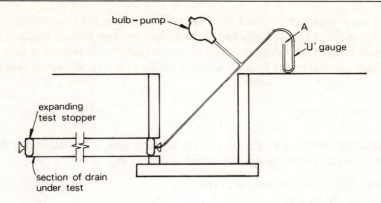

Fig. 8.3 Low-pressure air test

Whenever the bulb is pressed, the open end of the 'U' tube, marked 'A' in Fig. 8.3 must be sealed by finger pressure, or all the water will immediately spurt out. The finger may be released as soon as the valve in the pump closes.

The level of the water in the gauge should not drop more than 25 mm in five minutes, or 12.5 mm in five minutes in a system containing traps.

High pressure
An extremely rigorous test may be applied to a drain run using expanding stoppers in combination with a piston pump (hand or foot operated) and a pressure gauge. Care must be taken to avoid 'bursting' the pipe under test – in practice a rubber or plastic connection will usually fail before the pipework does, and having been released the excess pressure will probably reseal itself. Alternatively, one of the expanding stoppers will probably give way, releasing all pressure (and not resealing).

One of the most useful applications for a 'high-pressure' test is when used in conjunction with a smoke 'bomb' to locate leaks – a smoke test on its own can be a slow and incongruous affair, but the effects are often considerably accelerated by use of high-pressure equipment.

Although drains of all types can be tested at high pressure, the site engineer will occasionally come across a situation where a section of pipework is *intended to operate* at high pressure and therefore such tests are essential. Such situations are frequently found in chemical works, oil and gas terminals, water and sewage treatment works, pumping stations, etc. and the contract documents will specify both design and test pressures.

Smoke bomb

This is a kind of firework which produces dense, penetrating smoke. A smoke-bomb has its touch-paper fuse lit, and is immediately sealed inside a suspect drain run. Careful preparations should be made before lighting the fuse, to ensure that all pipe openings are sealed off, and that the last stopper goes in easily.

If smoke escapes, the drain leaks. The advantage of this system is that it can usually indicate where the leak is, whereas an air-pressure test alone cannot.

Soil investigations

Frequently the site engineer is expected to carry out, or assist with, subsoil investigations and from the data obtained decisions can be made on:

(a) Foundation depth, design and type
(b) Import of fill material
(c) Export of subsoil (for sale if marketable)
(d) Stabilisation of the subsoil
(e) De-watering requirements
(f) Plantings

The engineer must not, of course, be expected to provide specific recommendations in all or any of the above categories, but he must be able to provide the specialists who can, with the information they need. A site engineer is most often involved with 'initial' or 'preliminary' investigation, probably even before consultants are appointed. It is essential that any information obtained is accurately stated, and coherently recorded – in standardised terms that are immediately familiar to those subsequently involved in their interpretation.

Dig a hole

The most effective way to find out about subsoil structure is by excavating trial pits at random intervals across the site, and carefully inspecting the exposed material. Water-table levels can be rapidly determined in this way, provided some time is given for the pit to fill naturally. Soil samples may be taken at intervals, and stored in airtight, screwtop specimen jars – each must be accurately labelled with the following basic information:

(a) Site name
(b) Date
(c) Trial pit number (also accurately marked on a site plan)
(d) Depth at which sample taken

Each trial excavation should have its own diagrammatic record sheet, and a typical example is shown in Fig 8.4. A typical set of soil identification symbols is shown in Fig. 8.5.

At the conclusion of the site engineer's preliminary investigations, all samples and record sheets may be sent to a soils laboratory for analysis and recommendations (probably in conjunction with a consulting engineer) and it is essential at this stage for the specialists to have information on the above-ground proposals

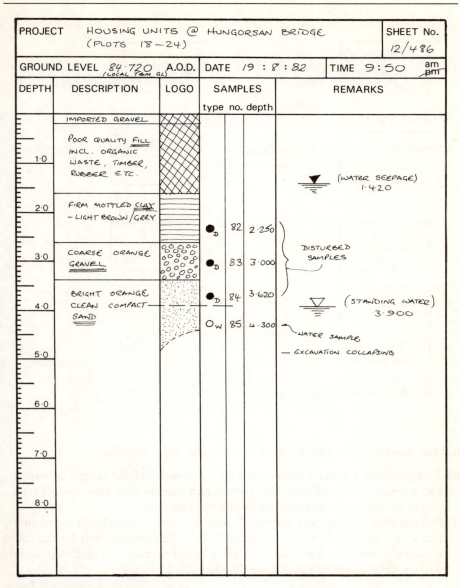

PROJECT	HOUSING UNITS @ HUNGORSAN BRIDGE (PLOTS 18-24)				SHEET No. 12/486

GROUND LEVEL 84·720 A.O.D. (LOCAL PbM GL) DATE 19 : 8 : 82 TIME 9:50 am/pm

DEPTH	DESCRIPTION	LOGO	SAMPLES type no. depth			REMARKS

Depth markings: 1·0, 2·0, 3·0, 4·0, 5·0, 6·0, 7·0, 8·0

IMPORTED GRAVEL

POOR QUALITY FILL INCL. ORGANIC WASTE, TIMBER, RUBBER ETC.

FIRM MOTTLED CLAY – LIGHT BROWN/GREY

COARSE ORANGE GRAVEL

BRIGHT ORANGE CLEAN COMPACT SAND

D 82 2·250

D 83 3·000

D 84 3·620

O W 85 4·300

(WATER SEEPAGE) 1·420

DISTURBED SAMPLES

(STANDING WATER) 3·900

WATER SAMPLE

EXCAVATION COLLAPSING

Fig. 8.4 Trial excavation record sheet

for the site, so that only the most appropriate tests and recommendations are considered.

Field identification and classification

There are two main reasons why the site engineer should attempt a basic 'classi-

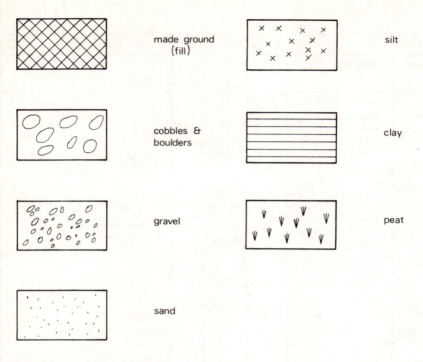

Fig. 8.5 Soil identification symbols

fication' description in the standard terms of the soils specialists,

(a) This provides a general cross-check on the identity of the sample provided, but a more subtly different description and sample may alert the soils engineer to conditions not noticed by the site engineer

(b) Soil samples are generally very small, and may not be completely representative – they are also (usually) 'disturbed'. It is therefore difficult for the soils engineer to imagine the exact conditions found on site, and an independent written description provides a useful cross-check

A very simple series of field tests is shown in Table 8.1. These are crude, almost non-scientific – and yet, carefully applied, they are usually enough for the site engineer to determine the general conditions prevailing, and to prepare a useful set of samples for further analysis in a soils laboratory. The loading values given in the table are approximate safe bearings, assuming distributed loads on a conventional 600 × 225 mm concrete strip foundation.

A more comprehensive set of field tests and site identification techniques may be found in the Building Research Establishment Digest No. 64 and BS 1377.

Table 8.1 Field identification and classification (based on table to Reg. D7, *Building Regulations* 1976)

Subsoil type	Subsoil condition	Field test	Approximate safe distributed bearing value in 600 × 225 mm concrete strip foundation (kN/m²)
Rock	Not inferior to limestone, sandstone or firm chalk	Requires at least a pneumatic or other mechanically operated pick for excavation	(unlikely to require conventional foundation)
Gravel, sand	Compact	Requires pick for excavation. Wooden peg 50 mm square hard to drive beyond 150 mm	100.00
Clay, sandy clay	Stiff	Cannot be moulded with the fingers and requires a pick or mechanically operated spade for its removal	100.00
Clay, sandy clay	Firm	Can be moulded by substantial pressure with the fingers and can be excavated with graft or spade	83.33
Sand, silty sand, clayey sand	Loose	Can be excavated with a spade. Wooden peg 50 mm square can be easily driven	50.00
Silt, clay, sandy clay, silty clay	Soft	Fairly easily moulded in the fingers and readily excavated	46.00
Silt, clay, sandy clay, silty clay	Very soft	Natural sample in winter conditions exudes between fingers when squeezed in fist	33.33

Standard penetration test

This is a useful method of directly measuring the penetration resistance of subsoil on the site, and of obtaining soil samples from considerable depth. A specialist contractor is usually employed for this operation.

The equipment consists of:

(a) Large tripod
(b) Two- or four-stroke winch drive motor
(c) Auger boring tool and rods

Tests and investigations

(d) Split-barrel sampler
(e) Sampler rods
(f) Drive assembly

Standard penetration tests (SPTs) might be instructed at, for instance, 1 m
below ground level and then at 2 m intervals to a total of 15 m below ground; a
total of eight tests.

Procedure

A hole is first prepared to the correct depth using the power auger, and then the
drive assembly, split-barrel sampler and rods are set up such that the sampler is in
position at the bottom of the borehole, and there is a free drop of 760 mm
available in the drive assembly.

The sampler is then 'seated' at the bottom of the borehole by being driven 150
mm into the soil with blows from the 65 kg drive hammer falling 760 mm. The
hammer is lifted up its shaft after each blow by a cable pulled by the winch drive
motor. The number of blows needed to 'seat' the sampler 150 mm are recorded.

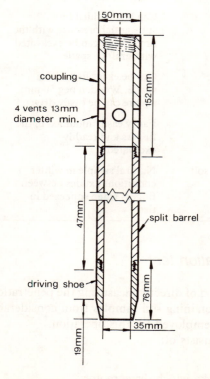

Fig. 8.6 Standard penetration sampler

The sampler is then driven a further 300 mm by the same process, stopping at every 75 mm of penetration to record the number of blows supplied to that point. If fifty blows have been applied (not including the 'seating' drive) and the penetration depth is still less than 300 mm, the test is discontinued and the *actual* penetration measured.

The total number of blows required for the 300 mm penetration below the 'seating' drive is termed the penetration resistance (N).

The sampler is then withdrawn, and its contents placed in a fully labelled sample jar for analysis by a soils laboratory.

Each SPT should have the following details recorded:

1. Penetration resistance (N_1 the number of blows)
2. Start/finish depths of the test
3. Number of blows for initial 'seating' 150 mm
4. Number of blows and depth of penetration if less than 300 mm
5. Date of test
6. Identification number and location of borehole
7. Reduced level of ground at surface (AOD)
8. Boring method
9. Data on water levels in the borehole
10. Soil description and sample particulars

Standard penetration tests are usually carried out in cohesive soils (clays, silts), but for non-cohesive soils (sands, gravels) a solid 60° core may be used instead of the open-ended sampler – this does mean that the test will produce no final sample for analysis.

A typical split-barrel sampler assembly is shown in Fig. 8.6.

Chapter 9

Excavation plant and labour

The site engineer is interested in machines and men in terms of output. This is not to say that he does not also appreciate other qualities of both in various ways, but professionally he is limited to a consideration purely of effectiveness in terms of cubic metres per day.

Machines versus men

In recent decades the use of large gangs of labourers equipped with hand-tools and engaged simply in earthmoving, has declined significantly. The mechanical excavator has almost completely surpassed *Homo sapiens* as a digging machine, except for small or very specialised works. The main reason for this is money – it often costs less per cubic metre to excavate by machine, provided that there is enough work of the right kind for a machine to do. The site engineer should not, however, ignore hand excavation as a possible resource. An enthusiastic gang of well-equipped groundworks labourers can make short work of a new bungalow foundation or an individual drain run, often with greater precision and less mess than is possible by mechanical excavation. For short-term precise operations of this kind, 'hand-dig' can be competitive in terms of cost, provided that conditions are suitable.

As a general guide, housing sites of up to twenty units are more economically dealt with by hand excavation. Over this level it usually becomes cheaper to use mechanical excavators. In practice, the engineer will often be able to justify mechanical excavation on much smaller sites, on the basis of time-saving, elsewhere in the contract programme.

Machines

There are many types of excavating machine, and a complete survey would require a book of its own. The very large machines can move incredible volumes of material, and their capacity for work can be enormous – they are expensive to hire, but in terms of output the unit cost per cubic metre can be very low under good conditions. In circumstances where the engineer has a free hand, the largest capacity machine available should be used, provided that there is continuity of work.

Although the site engineer will sometimes come into contact with the very large scrapers, graders, drag-lines, etc. he will rarely have any direct control over the selection of these items, nor have any day-to-day responsibility for their use on site, other than the provision of setting-out data.

On the other hand, the smaller excavators are commonly under the direct control of the site engineer and foremen for long periods and therefore of considerably more concern in terms of performance characteristics and capacities. The two most popular types of small excavator are:

(a) The 'wheel digger'
(b) The '360° tracked digger'

Machines of these basic configurations are common to almost all kinds of construction site, and collectively do most of the world's earth-moving. It is appropriate here to limit consideration to one typical example of each pattern, and to refer the site engineer to specialist works on mechanical plant, the trade journals and the manufacturers, for data on the larger machines as and when required.

'Wheel digger'

Originally based on a farm-type tractor with large wheels at the rear and much smaller ones at the front, this type of machine is very popular, principally because it can be easily driven around on site, and can (properly licensed and insured) be driven on public roads, at speeds up to 29 km/h. Machines of this type generally have a loading bucket at the front and an excavating bucket at the rear. The excavating bucket (the whole assembly is referred to as the 'back actor') is typically used for foundation or drain trench excavation. The loading-bucket arrangement (referred to as the 'shovel') is often used for loading loosened spoil into lorries for disposal, although it can also be used for heavy excavation, site stripping, etc, if ground conditions are such that it can obtain suitable traction. Four-wheel-drive versions of these machines are available, and this feature can improve site performance in wet conditions.

In almost all variations of this machine, the driver sits on a swivel seat enabling him to face forward for manoeuvring and loading, and rearwards for excavating trenches, etc. The operator is protected from the elements by a fully glazed cab, giving maximum visibility, and should wear some form of ear protection against continuous engine noise.

Figure 9.1 shows the operating range for a typical machine in this class.

The main disadvantages of this type of machine are:

(a) Relatively poor traction in wet or boggy terrain
(b) Limited excavation/loading area; 180° arc at rear of machine, and limited by boom length

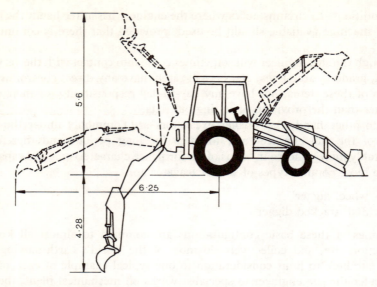

Fig. 9.1 Wheel digger

To excavate a foundation trench, the machine is reversed into a position such that the centre-line of dig falls somewhere between the two rear wheels. All four wheels are then lifted clear of the ground, the front ones by making the 'shovel' thrust downwards and the rear ones by operating two hydraulic jacks situated at each end of the kingpost track at the rear. The kingpost is then moved sideways by hydraulic rams until it is exactly over the dig centre-line, and the machine is now ready to 'pull' a trench.

The three most generally useful excavator bucket sizes are:

(a) 350 mm (gas/water/telephone/electricity service trenches, land drain-laying)
(b) 650 mm (standard concrete strip foundations)
(c) 950 mm (bulk excavation)

Various alternative buckets and fittings are available for specialised functions, and these include:

1. Ditching tool
2. Asphalt cutter
3. Ripper tooth
4. Lifting hook
5. Power breaker tool
6. Hydraulic grab

'360° digger'

Based on an undercarriage similar to that employed in a military tank, the

138

superstructure of a machine of this type can be rotated through a full 360°. This means that the machine can excavate at full stretch in one direction and slew round through a full circle to load a lorry at full stretch in any other direction. In the case of the 3050 mm dipper arm machine shown in Fig. 9.2 the maximum distance between excavation point and disposal lorry can be as much as 15 m, without the machine moving its tracks. It also means that a house or bungalow foundation can often be almost completely excavated from one position, but despite the considerable saving in time, this should be avoided whenever possible as the finished effect can be a very untidy foundation excavation, and a very expensive one to concrete. The undercarriage tracks ensure immense grip, and also help to distribute the machine's weight very evenly, so that this type of excavator can operate effectively in very soft conditions – even partly submerged in water if the need arises. A comparison of manufacturer's data is given in Table 9.1.

Trench excavator

The tracks must be centred exactly over and parallel with the centre-line of dig. This requires considerable skill as precise manoeuvring with a track-steered machine is a difficult operation. No lifting or bracing is required, and moving along the line of the trench as work proceeds is a simple matter of forward (or

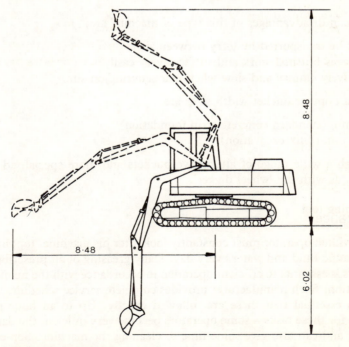

Fig. 9.2 Tracked digger

Table 9.1 Comparison of manufacturers' data (all dimensions in metres unless stated)

	JCB 3CX	Hymac 580D (3050 mm dipper arm)
A. Max. dig depth	4.28	6.02
B. Max. reach (GL)	6.25	8.48
C. Max. working height	5.6	8.48
Engine power	54 kW	82 kW
Max. forward speed	29.2 km/h	2.4 km/h
Breakout – dipper	26 kN	41.0 kN/57.0 kN†
– bucket	40 kN	70.0 kN
Weight	6.43 tonnes	12 tonnes*
Max. excavator bucket	0.30 m³	0.59 m³
Fuel tank	90 litres	270 litres

*Manufacturer's figure for standard machine with 1800 mm dipper arm
†Optional hard-dig position on dipper

reverse) travel. Steering is executed by varying track speeds.

Much damage can be caused by the tracks to raised kerbs, and also to flat concrete or blacktopped areas. This is often a very good reason for not using such a machine, particularly where reinstatement costs outweigh any time savings made.

The main disadvantages of this type of machine are:

(a) Must be transported by lorry between sites
(b) Damages finished work and surfaces very easily
(c) Relatively clumsy and slow when manoeuvring on site.

The most popular bucket width sizes are:

(a) 600 mm (standard concrete strip foundation)
(b) 1200 mm (bulk excavation)

– although a wide range of alternative buckets and more specialised fittings is available, as with the 'wheel digger'.

Maintenance

The individual operator must constantly look after his machine, topping up fuel, oil, hydraulic fluid and water every day. Daily greasing of all prescribed nipples and joints is essential to efficient operation in accordance with the manufacturer's specification. Each manufacturer provides complete service schedules and charts and it is essential that these are followed exactly. Up to an hour per day is allocated for these tasks – some operators become very quick at the daily routine servicing and can also give some time to cleaning the machine. Some operators will visit the site during the weekend to clean and polish their machines.

Constant checks should be made for leaks and spills, and immediate investigations *must* be made when any sudden decline in operating performance is noted. The site engineer must ensure that any 'down-time' is recorded accurately and proper deductions made from plant time-sheets. Excavation plant is very expensive and hiring rates take breakdowns and repairs into account. It is usual to consider that all plant is 'hired' to the contract even if owned outright by the main contractor, because this puts its real cost into perspective. Some major repairs can be avoided completely by careful maintenance and use, others are a direct consequence of the number of operating hours amassed.

Operators

'Digger drivers' are almost always self-taught. Just as some people have a natural affinity for driving cars or an innate ability to fly aircraft, others have a native flair for handling complex excavation equipment. To an operator whose aptitude is instinctive, a mechanical excavator is just an extension of the various extremities of his body – he thinks of the bucket as his hand, the boom as his arm and the chassis as his pelvis and legs. Confined to his cab for most of the working day, isolated by noise and the workings of his machine, operators will sometimes become angry or frustrated over a comparatively trivial matter and remonstrate with supervisory staff. It is very rare, however much he feels he has been provoked, that an operator will damage his equipment or the work he has done – and almost without precedent that he will carry out threats to 'bury that building inspector up to his neck' (etc.), however meaningful such statements may sound at the time.

Safety and security

Excavators should always be 'secured' when left unattended – even if left for a few minutes the ignition key should be removed, and under no circumstances should an engine be left running. To prevent an accident in the case of failure in the hydraulics system, or unauthorised tampering, the booms should be lowered so that the bucket rests firmly on the ground.

When left on site overnight, all excavators and any other movable plant should be locked in a security compound under constant lighting.

Left on site over a weekend or public holiday machines can be temporarily disabled by having their batteries removed – this is usually sufficient to discourage the casual 'borrower' – and it is sometimes helpful to leave them all on charge in an unmarked shed ready for a new week's operation. 'Permanent' disabling can be arranged with a plant fitter, and will probably involve removal of some vital part of an engine or transmission. This should not be attempted without expert advice.

Vandalism

The popular target for vandals is usually the glass in cab windows and lighting fixtures. Perhaps the easiest way to prevent damage is to park the machine sufficiently distant from the compound fence, such that missiles simply will not reach – alternatively any windows at risk can be boarded over, and light fittings wrapped in dust sheets to disguise their shape and cushion any impact. Tank caps should be lockable and *locked*. Doors, tool and engine compartments should be locked wherever possible. Such simple security precautions should take only a few minutes to execute.

Mechanical excavation rates

The rate at which spoil may be excavated, for any given machine, depends on:

1. Type of soil
2. Topography of ground
3. Operator efficiency
4. Disposal capacity (lorry sizes, tip distances)

– and consideration of each of these variables will cause the engineer to suspect any definitive published data on the subject.

The site engineer will, however, find himself in a position where he must obtain a reasonably accurate output figure for a particular machine. Almost the only way to determine this accurately is to have the machine dig a test volume, and time the result. Multiplied over the whole working day, with allowances for meals and discontinuities, an output figure can be assessed and the likely cost per unit predicted.

A comparison of men and machine hours is given in Table 9.2 and some quite useful excavation rates can be obtained from this for purposes of approximate estimates.

Under unrestricted and ideal conditions of mass excavation a tracked excavator of the type shown in Fig. 9.2 can excavate at a rate of more than 1000 m_3 per day, but as can be seen from Table 9.2 this can be reduced to less than 10 m^3 per day in adverse conditions.

Labour

'Hand-dig' is most often employed for precision operations, where the volume to be excavated is relatively small and the task is comparatively complex. A good example of this is the installation of a service pipe or drain run across an existing footpath or roadway – a mechanical excavator would almost inevitably damage telephone, electrical, gas and water services, whereas a man excavating in the same area is unlikely to do so.

Table 9.2 A comparison of machine/man hours for excavation

Operation	Hours per cubic metre	
	Man	Machine
Excavate basement n.e. 1 m deep		
Loam, or light sandy soil	3.00	0.133
Heavy clay	3.75	0.177
Chalk	7.50	0.266
Hard rock	15.00	0.665
Excavate trench n.e. 1.5 m deep		
Loam, or light sandy soil	3.25	0.17
Heavy clay	4.06	0.23
Chalk	8.13	0.34
Hard rock	16.25	0.85
Excavate trench 1.5–3.0 m deep		
Loam or light sandy soil	5.00	0.22
Heavy clay	6.25	0.29
Chalk	12.50	0.44
Hard rock	25.00	1.10

Note: These rates are for illustration purposes only and should be varied to take account of site conditions.

The 'labourer'

Hand excavation is almost always carried out by an employee called a 'labourer' – this is a term which to some people implies a lack of skill or expertise, but whilst it is true that labourers may lack the specialist technical training obtained by qualified tradesmen they must nevertheless possess considerable skill in performing their own distinct function. Much damage can be done to the human frame by incorrect lifting and levering – and it requires much more than brute force to endure a full week of careful hand-dug trenching work.

Many labourers are good at drainlaying or concreting – some develop unusual skills such as preparing 'benching' to manholes, in confined spaces and conditions where no ordinary plasterer or bricklayer would venture. A completely unskilled individual has no place on a construction site, and the term 'labourer' should never be used in isolation, but should properly be prefaced by the type of work carried out, e.g. 'groundworks labourer', 'brickworks labourer', etc.

Excavation tools

The most useful general-purpose tool is a shovel, but for clay and clay-bound soils a crescent-shaped spade called a 'graft' can be used – although sometimes a fork does just as well in this situation. For granular soil a pick is used to break up compacted material, and then a shovel is used to clear loosened debris.

A wheelbarrow is essential to take spoil from the point of excavation to a disposal heap (if not to be backfilled), and a collection of scaffold boards should be provided to make a continuous walkway over soft ground. The scaffold boards must not be those used for superstructure scaffolding, and should be clearly identified for their purpose.

Index